ELECTRICITY, MAGNETISM, AND ANIMAL MAGNETISM

A CHECKLIST
of
Printed Sources

1600-1850

EXPERIMENTS

AND

OBSERVATIONS

ON

ELECTRICITY,

MADE AT

Philadelphia in *America,*

BY

Mr. BENJAMIN FRANKLIN,

AND

Communicated in several Letters to Mr. P. COLLINSON,
of *London,* F. R. S.

LONDON:

Printed and sold by E. CAVE, at *St. John's Gate.* 1751.
(Price 2*s.* 6*d.*)

ELECTRICITY, MAGNETISM, AND ANIMAL MAGNETISM

A CHECKLIST
of
Printed Sources

1600-1850

Compiled by
ELLEN G. GARTRELL
for the
American Philosophical Society

SR Scholarly Resources Inc.
1508 Pennsylvania Avenue · Wilmington, Delaware 19806

Library of Congress Catalog Card Number: 75-29743
International Standard Book Number: 0-8420-2078-0

SCHOLARLY RESOURCES, INC.
1508 Pennsylvania Avenue
Wilmington, Delaware 19806

Printed in the United States of America

Communications concerning this publication should be
addressed to the American Philosophical Society Library,
105 South 5th Street, Philadelphia, Pennsylvania 19106.

Table of Contents

INTRODUCTION vii

SYMBOLS ix

ELECTRICITY AND MAGNETISM 1600-1850
    Before 1801 1
    1801-1850 49
    Periodicals 85

ANIMAL MAGNETISM
    Before 1801 87
    1801-1850 96
    Periodicals 105

ARTICLES ON ELECTRICITY AND MAGNETISM IN AMERICAN PHILOSOPHICAL
    SOCIETY *TRANSACTIONS* BEFORE 1850 107

MISCELLANEOUS: MAPS, BROADSIDES, MANUSCRIPT BOOKS 111

APPENDIX: BOOKS FROM BENJAMIN FRANKLIN'S LIBRARY AND NOTES ON
    PROVENANCE 113

SUBJECT AND AUTHOR INDEX 115

Introduction

This checklist originated as a project of the Library of the American Philosophical Society to make known its holdings in the history of electricity, magnetism, and animal magnetism. The Library was begun in the eighteenth century, a time of great interest in Natural Philosophy and especially in electricity as one of the newest branches of that study. Not surprisingly, therefore, a small collection of books on the subject accumulated, including some volumes from the library of the Society's founder Benjamin Franklin (see Appendix, p. 106). A printed catalogue of the Society's books in 1824 listed 45 titles under the heading Electricity and Magnetism. With increasing interest in our time in both the history of the science and in Franklin as one of its early practitioners, the collection is growing, and now numbers over 400. Franklin's involvement is also one basis of the Society's holdings in animal magnetism. In 1784 the King of France appointed him to a commission to investigate animal magnetism, or Mesmerism, which was the rage in Paris, and Franklin acquired a number of pamphlets about it, some of which the Society now owns.

As it became clear that other useful collections on early electricity, magnetism, and animal magnetism exist in Philadelphia, it was decided to augment this guide by including the holdings of five other libraries: the Library Company of Philadelphia, the College of Physicians of Philadelphia, the libraries of the University of Pennsylvania, the Franklin Institute, and the Historical Society of Pennsylvania. These libraries house the largest electrical collections in the center-city area; others in Philadelphia which have additional books and pamphlets on electricity are mentioned at the end of this Introduction.

Titles were collected by searching card catalogues and shelves and by checking lists in several standard electricity reference books (Ronalds & Wheeler Gift, see p. viii) with the Union Library Catalogue of Pennsylvania. The primary goal was to locate and include all possible different titles and editions; somewhat less attention was devoted to finding all copies of each. The holdings of the American Philosophical Society Library are probably the most completely represented in this guide, because it was possible to spend the greatest amount of time there.

Certain categories of general works, such as early encyclopedias and learned societies' serial publications, which often contain important articles on electricity, have been omitted altogether, although these works are held in great numbers by the libraries surveyed. Some broad treatises and textbooks on natural philosophy have been included when they contain large sections relevant to electricity or magnetism. Post-1850 reprints of early works are not included, as these are usually readily available. About 100 of the titles listed are reprints or extracts from periodicals; these are identified as to source and are interfiled in the main lists. An exception is made for 36 articles from the APS *Transactions* before 1850, which form a separate section, Nos. 1243 to 1279. The entire guide includes about 1300 entries.

The titles have been numbered in one series. The two broad groupings "Electricity and Magnetism 1600-1850" and "Animal Magnetism to 1850" are each divided into two periods: "Before 1801" and "1801-1850." Works of authors whose writings spanned both time periods appear together in the earlier section; cross references and an index have been used to locate additional related titles. Selected subject headings are included in the index to guide the user to major sources for certain topics. Each title, except those on animal magnetism, has been checked in the "Ronalds," "Mottelay," and "Wheeler Gift" catalogues (see p. viii for complete citations) and its presence or absence in these standard works is noted by symbols described on p. ix, where library location symbols are also explained.

No attempt has been made at complete bibliographic description. Titles have been shortened when this could be done without confusion of editions or loss of information as to content. The publisher's name is given only when needed to clarify different editions. Places of imprint have been anglicized whenever possible. Annotations have been kept to a minimum. Provenance is mentioned in selected cases (see Appendix, p. 106).

The cutoff date of 1850 was chosen for this list mainly to prevent its becoming unmanageably long. Nearly all of the six libraries are rich in later nineteenth century works, though with differing emphases. Especially noteworthy are the electrotherapeutics collection at the College of Physicians and the electrical technology books at the Franklin Institute, where a special effort was made to collect electrical books for the International Electrical Exhibition held there in 1884.

This guide to electricity resources is only a beginning. Most of the libraries, including the Philosophical Society, continue to add to their collections in the field; other libraries in the Philadelphia area hold additional volumes valuable to electrical researchers. Among these are the Free Library of Philadelphia, Pennsylvania Hospital, the Wagner Free Institute of Science, and—a little farther afield—the Academy of the New Church, Bryn Athyn, Pa. and Lehigh University, Bethlehem, Pa.

This project was undertaken on a grant from the Andrew W. Mellon Foundation to the American Philosophical Society. I would like to express my thanks to all the participating libraries and the many individuals in each who aided in the search for elusive electricity. I would like to thank especially Dr. Rudolf Hirsch of the University of Pennsylvania and Mrs. Lisabeth M. Holloway of the College of Physicians for their generous contributions of time, space, and ideas. Dr. Whitfield J. Bell, Jr. and Roy E. Goodman at the Philosophical Society were constant sources of facts and guidance. I am very grateful to Mrs. Jean T. Williams for her care and attention in typing the manuscript.

E. G. G.

References

Mottelay, Paul Fleury. *Bibliographical History of Electricity and Magnetism Chronologically Arranged . . .* London: Charles Griffin & Co., 1922.

Ronalds, Sir Francis. *Catalogue of Books and Papers Relating to Electricity, Magnetism, the Electric Telegraph, &c. Including the Ronalds Library.* London: E. & F. N. Spon, 1800.

Weaver, William D., ed. *Catalogue of the Wheeler Gift of Books, Pamphlets, and Periodicals in the Library of the American Institute of Electrical Engineers.* New York: American Institute of Electrical Engineers, 1909.

Symbols

Reference symbols (see p. viii for complete citations)

M   =   This title is mentioned in Mottelay's *Bibliographical History* . . . Mp398 — In Mottelay page 398

R   =   In *Ronalds Library Catalogue*

W   =   In *Wheeler Gift Catalogue.* W47 — Catalogue No. 47

Not in MRW — Title not found in any of above three

(R) =   Title found in Ronalds, but not the same as copy in hand.
Also used for M and W, e.g. (Mp422), (W806)

Library Symbols

APS =   American Philosophical Society Library

LCP =   Library Company of Philadelphia

HSP =   Historical Society of Pennsylvania

CP   =   College of Physicians of Philadelphia

FI   =   Franklin Institute

UP   =   University of Pennsylvania

ELECTRICITY & MAGNETISM 1600-1850

Before 1801

1. ADAMS, GEORGE, 1750-1795.
Essay on electricity; in which
the theory and practice of that
useful science, are illustrated by
a variety of experiments, arranged
in a methodical manner. To which
is added, an essay on magnetism.
xvi, 367p. 6pl. London, 1784.
Mp281; R; W519.   LCP

2. ----. An essay on electricity explaining the theory and
practice of that useful science;
and the mode of applying it to
medical purposes; with an essay on
magnetism... 2d ed. x, 476,
[18]p. plates, front. London,
1785.
Ms. notes: last 4p.
Mp281; R.   LCP, APS

3. ----. An essay on electricity...with an essay on magnetism. 3rd ed. 473p. 7pl.
London, 1787.
Mp566; R.   FI, UP, CP

4. ----. An essay on electricity, explaining the principles of that useful science; and
describing the instruments...to
which is now added a letter to the
author from Mr. John Birch, sur-
geon, on the subject of medical
electricity. 4th ed. xi, 588,
15p. 5pl., front. London, 1792.
Mp280; R; W519a.   FI, UP, LCP,
                          APS

5. ----. Lectures on natural and
experimental philosophy,...
2d ed. with considerable corrections and additions, by William

Jones. 4v. & lv. of plates.
London, 1799.
Appendix to v.1 bound separately.
Not in MRW.   APS

6. [----]. A summary view of the
general principles of electricity. 76p. plate. [London,
1784?].
No t.p.
Not in MRW.   APS

7. [ADANSON, MICHEL], 1727-1806.
Lettre du Duc de Noya Carafa
sur la tourmaline, à Monsieur de
Buffon. 35p. lpl. Paris, 1759.
Benjamin Franklin's copy (HSP).
Mp193; R.   HSP, APS

8. AEPINUS, FRANÇOIS ULRIC THEODOR,
1724-1802. Tentamen theoriae
electricitatis et magnetismi. Ac-
cedunt dissertationes duae, quarum
prior, phaenomenon quoddam electri-
cum, altera, magneticum, explicat.
[21], 390p. 7pl. St. Petersburg,
[1759].
Autograph: Benjamin Franklin.
Mp217; R; W395.   APS

see also: 507

9. AHLWARDT, PETER.
Bronto-theologie; oder, vernünftige und theologische Betrachtungen
über den Blitz und Donner... 410p.
illus. Greifswald, 1745.
Not in MRW.   FI

ALDINI, GIOVANNI: see 195, 197

10. [ALENCÉ, JOACHIM D'], d. 1707.
Traitté de l'aiman; divisé en

Figure 1. Willem van Barneveld, *Geneeskundige Electricity.*

(See No. 23)

deux parties. La première contient
les expériences & la seconde les
raisons que l'on en peut rendre.
Par Mr. D---. [20], 140, [8]p.
33pl., front. Amsterdam, 1687.
Ms. notes (APS).
Mp554; R; W200.   FI, APS

11. ANDRY, CHARLES LOUIS FRANÇOIS,
     1741-1829. Rapport sur les
aimans présentés par M. l'Abbé Le
Noble... 15p. Paris, [1783].
From Registres de la Soc. royale
de médicine. No t.p.
Benjamin Franklin's copy.
R.   APS

12. ANTHEAULME.
     Mémoire sur les aimants arti-
ficiels, qui a remporté le prix de
physique de l'académie des sciences de S. Pétersbourg,... 59,
[3]p. plate. Paris, 1760.
Benjamin Franklin's copy (HSP).
Mp274; R.   HSP, APS

see also: 366

ANTINORI, VINCENZIO: see 534

13. BAMMACARUS, NICOLAUS, d.
     ca. 1778. Tentamen de vi
electrica ejusque phaenomenis, in
quo aeris cum corporibus universi
aequilibrium proponitur... [4] x
[2, 8], 202 [4]p. Naples, 1748.
Mp273; R.   APS

14. BARBERET, DENYS, 1714-1770.
     Dissertation sur le rapport
qui se trouve entre le phénomènes
du tonnerre et ceux de l'électricité,... 16p. Bordeaux, 1750.
Mp167, 321; R.   APS

BARBEU-DUBOURG: see 181

15. [BARBIER DE TINAN].
     Memoire sur la maniere d'armer
d'un conducteur la cathédrale de
Strasbourg et sa tour. 34p.
[Paris, 1780].
Inscription by author to Benjamin
Franklin.
R.   APS

16. ----. Nuove considerazioni
     sopra i conduttori...tr. dal
francese. 43p. plates. Venice,
[1779?].
R.   LCP

see also: 520

17. BARKER, FRANCIS, d. 1859.
     ...De invento Galvani, vulgo
animalium electricitate dicto...
41p. Edinburgh, 1795.
Inaugural dissertation.
Autograph: James Rush.
(R).   LCP

18. [BARLETTI, CARLO], d. 1800.
     Analisi d'un nuovo fenomeno
del fulmine ed osservazioni sopra
gli usi medici della elettricità.
[8], 63p. 1pl. Pavia, 1780.
R.   LCP, CP, APS

19. ----. Dubbi e pensieri sopra
     la teoria degli elettrici
fenomeni. xxvii, 136p. 1pl.
Milan, 1776.
Contains a letter to Volta p.118.
R.   LCP

20. ----. Nuove sperienze elet-
     triche secondo la teoria del
Sig. Franklin e le produzioni del
P. Beccaria... 134p. 1pl. Milan,
1771.
Mp207; R; W431.   FI, LCP, APS

21. ----. Physica specimina.
     [7], 184p. 2pl. Milan, 1772.
R.   APS

22. [BARLOW, WILLIAM], d. 1625.
     Magneticall advertisements:
or Divers pertinent observations,
and approved experiments concerning
the nature and properties of the
loadstone:... [8], 86, [2]p. illus.
London, 1616.
Mp573; R; W89.   UP

see also: 461

23. BARNEVELD, WILLEM VAN, 1747-
     1826. Geneeskundige electri-
citeit. [3 parts]. xvi, 387,

4

[21], viii, 208, [12]p. 3pl.
Amsterdam, 1785.
R.   APS

24.   BAUER, FULGENTIUS, 1731-1832.
      ...De electricitatis theoria
et usu. 144p. Vienna & Leipzig,
1767.
Inaugural dissertation.
(R).   APS

25.   ----.   Experimental-Abhand-
      lung von der Theorie und dem
Nutzen der Electricität...und von
der Würkung der Luft-Electricität
in dem menschlichen Körper von
Marherr und Kirchvogel. 304p.
Chur & Lindau, 1770.
R.   APS

26.   [BAZIN, GILLES AUGUSTIN],
      1681-1754. Description des
courants magnétiques dessinés et
gravés d'aprés nature en XV plan-
ches, suivie de quelques observations sur l'aiman, par Mr.----...
[4], 52 [1], 23p. 20pl. Strasbourg, 1753.
Includes Supplément...1754.
Mp208, 273; R; W374.   APS

27.   ----.   Supplément pour la
      description des courants
magnetiques. 23p. 5pl. [Stras-
bourg, 1754].
W374.   APS

28.   BECCARIA, GIOVANNI BATTISTA,
      1716-1781. A sua altezza
reale il signor Duca di York
Sperienze, ed osservazioni. [Let-
ter] 16p. Turin, 1764.
R.   APS

29.   ----.   De atmosphaera elec-
      trica...Ad regiam Londinen-
sem Societatem libellus. 8p.
Turin, 1769.
Benjamin Franklin's copy.
R.   APS

30.   ----.   De' fiori elettrici...
      Lettera...al...Tiberio Caval-
lo. [9]-14p. [Milan, 1780].
Incomplete.

From Opusculi scelti sulle scienze
e sulle arti, III.
R.   APS

31.   ----.   Dell'elettricismo.
      Lettere...dirette al chiaris-
simo sig. Giacomo Bartolomeo Bec-
cari... [6], 378, [2]p. Bologna,
1758.
Autograph: "B. Franklin's sent him
by the Author 1760."
Mp207; R; W392bis.   APS

32.   ----.   Dell'elettricismo arti-
      ficiale, e naturale. Libri
due. [4], 245p. Turin, 1753.
R; W375.   APS, LCP

33.   ----.   Della elettricità ter-
      restre atmosferica a cielo
sereno; osservazioni... [3], 54p.
[Turin, 1775].
Benjamin Franklin's copy (APS).
Mp416; R; W450.   APS, LCP

34.   ----.   Elettricismo artificiale.
      viii, 439p. 11pl. Turin, 1772.
c.1: no plates; Benjamin Franklin's
copy (APS).
c.2: bookplate of George Sarton; no
t.p.   (APS).
Mp207; R; W435bis.   UP, LCP, APS

35.   ----.   Experimenta, atque
      observationes, quibus electri-
citas vindex late constituitur,
atque explicatur. [2], 66p. 2pl.
Turin, 1769.
Benjamin Franklin's copy (HSP).
Mp416; R; W424.   FI, HSP, APS

36.   ----.   Lettera...al...Gian-
      francesco Cigna...si sperimen-
ta nel voto intorno alla divergenza
elettrica de' pendoletti immersi
nell' olio... 19p. illus. n.p.
[1779].
Benjamin Franklin's copy.
Not in MRW.   APS

37.   ----.   Lettera...al...Gio.
      Francesco[?] Fromond sul
cangiamento di colore prodotto dal
fuoco; [with] Poscritta alla lettera del...Giambattista Beccaria

diretta al Signor Canonico Fromond.
10p. n.p. [1779].
Benjamin Franklin's copy.
R.   APS

38. ----. Lettre sur l'électri-
       cité, adressée à M. l'Abbé
Nollet. Tr. de l'Italien par M.
de Lor. viii, 144, [4]p. Paris,
1754.
R.   APS

39. ----. On terrestrial atmo-
       spheric electricity, during
serene weather... [tr. from the
Italian]. [421]-475p. [London,
1776].
With author's Treatise upon artificial electricity...
R; W457.   APS

40. ----. A treatise upon arti-
       ficial electricity...To which
is added an essay on the mild &
slow electricity which prevails in
the atmosphere during serene
weather. Tr. from the original
Italian... iv, [4], 475, [10]p.
11pl. London, 1776.
R; W457.   FI, LCP, APS

see also: 20, 146, 510, 740

41. BECK, DOMINIKUS, 1732-1791.
       Kurzer Entwurf der Lehre von
der Elektricität... [8], 196p.
8pl. Salzburg, 1787.
R.   APS

42. [BECKET, JOHN BRICE].
       An essay on electricity, con-
taining a series of experiments
introductory to the study of that
science;...intended chiefly with
a view of facilitating its application, and extending its utility
in medical purposes. 8, [x]-xv,
151p. Bristol, 1773.
Mp385; R; W438.   LCP, APS

43. [BELCHER, WILLIAM].
       Intellectual electricity,
novum organum of vision, and grand
mystic secret...being an experi-

mental and practical system of the
passions, metaphysics, and religion,
...with appropriate extracts from
Sir Isaac Newton, Dr. Hartley, Bed-
does, & others: with medical observations rising out of the subject.
By a rational mystic. [4], 184p.
London, [1798].
Not in MRW.   APS

44. BELGRADO, JACOPO, 1704-1789.
       I fenomeni elettrici con i
corollari da lor dedotti, e con i
fonti di ciò che rende malagevole
la ricerca del principio elettrico
... xii, 44p. Parma, 1749.
Mp555; R.   APS

45. BELLA, JOÃO ANTONIO DALLA,
       1730-1823. Noticias histori-
cas, e praticas acerca do modo de
defender os edificios dos estragos
dos raios... [14], 88p. 1pl.
Lisbon, 1773.
R; W439.   APS

46. BENNET, ABRAHAM, 1750-1799.
       New experiments on electri-
city, wherein the causes of thunder
and lightning as well as the constant state of positive or negative
electricity in the air or clouds,
are explained...Also a description
of a doubler of electricity, and of
the most sensible electrometer yet
constructed. With other new experiments and discoveries in the science
... [5], 141p. 3pl., front. Derby,
1789.
Bookplate: Silvanus P. Thompson (LC).
Mp289; R; W552.   LCP, APS

47. BÉRAUD, LAURENT, 1703-1777.
       Dissertation sur le rapport
qui se trouve entre la cause des
effets de l'aiman et celle des phenomenes de l'électricité,... 38p.
Bordeaux, 1748.
Mp164; R.   APS

48. ----. Theoria electricitatis.
       pp.95-144. [St. Petersburg,
1755].
With J. A. Euler's Disquisitio...

(q.v.)
Benjamin Franklin's copy.
W385.   HSP

49. ----. Theoria electricita-
        tis. 133-204p.
In Euler, J. Dissertationes selec-
tae...1757 (q.v.)
(R); (W385).   APS

50. BERDOE, MARMADUKE.
        An enquiry into the influence
of the electric-fluid, in the
structure and formation of ani-
mated beings. xxxii, 183p. 4pl.
(part colored). Bath, 1771.
Mp556; R; W432.   CP, FI

51. BERTHOLON, PIERRE, 1742-1800.
        De l'électricité des météores
... 2v. 6pl. Lyon, 1787.
Mp258, 295; R; W539.   FI, LCP,
                        APS

52. ----. De l'électricité des
        végétaux. Ouvrage dans lequel
on traite de l'électricité de
l'atmosphère sur les plantes, de
ses effets sur l'économie des
végétaux, de leurs vertus médico
et nutrivo-électriques et principalement des moyens de pratique
de l'appliquer utilement à l'agriculture, avec l'invention d'un
électro-végétomètre.
Mp258; (R); W512bis.   UP, LCP,
                        APS

53. ----. De l'électricité du
        corps humain dans l'état de
santé et de maladie;...dans lequel
on traite de l'électricité de l'
atmosphère, de son influence & de
ses effets sur l'économie animale,
&c. &c. xi, [1], 541p. Lyon,
1780.
Mp258; R.   LCP, APS

54. ----. De l'électricité du
        corps humain... 2v. 6pl.
Paris, 1786.
Mp129, etc.; R; W533   UP, CP, APS

55. ----. Mémoire sur un nouveau

moyen de se préserver de la
foudre. 30p. Montpellier, 1777.
Benjamin Franklin's copy.
R.   APS

56. ----. Nouvelles preuves de
        l'efficacité des para-ton-
nerres. 28p. 3pl. [Montpellier,
1783].
R.   LCP, APS

57. ----. Ueber die Elektricität,
        in Beziehung auf die Pflanzen;
die Mittel, die Elektricität zum
Nutzen der Pflanzen anzuwenden
u.s.w. Nebst der Erfindung eines
Elektro-vegeto-meters. x, [6],
301p. 3pl. Leipzig, 1785.
R.   LCP

58. [BETTI, LUIGI].
        L'Origine del Fulmine.
Poemetto [Dedicated to Franklin].
16p. Pisa, 1777.
Not in MRW.   APS

59. BIANCHINI, GIOVANNI FORTUNATO,
        1719-1779. Osservazioni in-
torno all'uso dell'elettricità
celeste e sopra l'origine del fiume
timavo riportate in due lettere.
[6], 92p. 1pl. Venice, 1754.
Probably from library of Benjamin
Rush.
R.   LCP

60. BINA, ANDREA, b. 1724.
        Eletricorum effectuum expli-
catio, quam ex principiis Newtoni-
anis deduxit novisque experimentis
ornavit. 157p. front. Padua,
1751.
R.   LCP, APS

BIRCH, JOHN: see 4

61. BLAGDEN, SIR CHARLES, 1748-
        1820. Proceedings relative to
the accident by lightning at Heck-
ingham... [by C. Blagden & Edward
Nairne]. 26p. 6pl. London, 1783.
Benjamin Franklin's copy.
W513.   HSP

62. BLUMENBACH, JOHANN FRIEDRICH,
     1752-1840. Elements of
physiology... Tr. from the original Latin, & interspersed with
occasional notes. By Charles Caldwell. To which is subjoined, by
the translator, an appendix, ex-
hibiting a brief & compendious
view of the existing discoveries
relative to the subject of animal
electricity. 2v. in 1. Philadelphia, 1795.
Not in MRW.   UP, CP

63. BÖCKMANN, JOHANN LORENZ, 1741-
     1802. Ueber Anwendung der
Electricität bei Kranken, nebst
der Beschreibung der neuen Maschine
von Nairne zur positiven und nega-
tiven Electricität, auch eines
neuen electrischen Bettes. 63p.
plate. Durlach, 1787.
Not in MRW.   APS

64. ----. Über die Blitzableiter.
     Eine Abhandlung auf höchsten
Befehl des Fürsten. 80p. Karls-
ruhe, [1791].
pp7 & 8 are photocopies (APS).
R; W568.   LCP, APS

65. BOETIUS DE BOODT, ANSELMUS,
     1550-1632. Gemmarum et lapi-
dum historia... Nunc vero recen-
suit, à mendis repurgavit, commentariis & pluribus, melioribusque
figuris illustravit, & multo locu-
pletiore indice auxit, Adrianus
Toll... [7], 576, [19]p. illus.
Leyden, 1636.
Mp17; (W120a).   APS

BOGGIA, JOSEPH: see 388.

66. BOHNENBERGER, GOTTLIEB CHRIS-
     TIAN, 1732-1807. Beschreibung
einer auf eine neue, sehr bequeme
Art eingerichtetem Elektrisir-
Maschine, nebst einer neuen Erfind-
ung, die elektrische Flaschen und
Batterien betreffend. 80p. 6pl.
Stuttgart, 1784.
(Mp434); R.   LCP, APS

67. ----. Beschreibung einer sehr
     wirksamen Elektrisir-Maschine
und einiger neuen elektrischen Versuche. Zweyte Fortsezung... [4],
138p. 4pl. Stuttgart, 1786.
R.   LCP

68. ----. Beschreibung einiger
     Elektrisirmaschinen und elek-
trischer Versuche. Dritte Fortsezung... [14], 224p. 5pl. Stuttgart, 1788.
R.   LCP

69. ----. Fortgesezte Beschreib-
     ung einer sehr Wirksamen Elek-
trisir-Maschine von ganz neuer
Erfindung und einiger zur elektrischen Praxis gehdrigen Werkzeuge
mit angehängten Versuchen. 110p.
6pl. Stuttgart, 1786.
R; W534.   LCP

70. BONCI CASUCCINI, ALESSANDRO.
     De luce et electricitate,
academica exercitatio, quam in
aula nob. Collegii Ptolemaei publice suscipiunt Alexander et An-
gelus Bonci Casuccini... [1], 19p.
1pl. Siena, 1792.
Not in MRW.   APS

BONCI CASUCCINI, ANGELO: see 70

71. BOND, HENRY, fl. 1675.
     The longitude found; or, A
treatise shewing an easie and
speedy way, as well by night as
by day, having but the latitude
of the place, and the inclination
of the magnetical inclinatorie
needle. [12], 65p. 7pl. [London,
1676].
No t.p.
M p118; R; W179.   APS

72. BONNEFOY, JEAN BAPTISTE, 1756-
     1790. De l'application de
l'électricité à l'art de guérir...
[2], 163p. Lyon, 1782.
Inaugural dissertation.
Mp299, 385; R.   APS

8

73. BORGOGNINI, ANTONIO MARIA.
    La teoria del fuoco, poema in
verso sciolto diviso in tre parti,
colle annotazioni e rami allusivi
d'un filosofo amico dell' autore
[Francesco M. Soldini]. 238,
[1]p. 3pl. port. [Florence, 1774].
Not in MRW.   APS

74. BOSE, GEORG MATHIAS, 1710-
    1761. De attractione et
electricitate; oratio inauguralis
... 38p. Wittenberg, [1739].
R.   APS

75. ----. Recherches sur la cause
    et sur la veritable téorie de
l'électricité. 56p. Wittenberg,
1745.
Benjamin Franklin's copy.
R.   HSP

76. ----. Tentamina electrica...
    vi, [2], 96p. Wittenberg,
1744. 3 commentaries: 1. De attractione et electricitate, 44p.,
1738.
2. De electricitate, 27p., 1743.
3. De electricitate inflammante
et beatificante, 24p., 1744.
R; W310.   FI, APS

77. ----. Tentamina electrica
    tandem aliquando hydraulicae
chymiae et vegetabilibus utilia.
[8], 50p. Wittenberg, 1747.
Commentaries 4 (27p.) & 5 (23p.).
R; W310.   APS

BOTTI, GIOVANNI BATTISTA: see 138

78. BOUGUER, PIERRE, 1698-1758.
    De la méthode d'observer en
mer la déclinaison de la boussole... [2], 7, 67p. 2pl. Paris,
1731.
R; W273.   APS

79. BOULANGER, NICOLAS ANTOINE,
    1722-1757. Traité de la
cause et des phenomenes de l'élec-
tricité. 2 parts in 1v. 2pl.
Paris, 1750.
Ms. notes.
Mp191; R; W356.   APS

80. BOYLE, ROBERT, 1627-1691.
    Experimenta et observationes
physicae: wherein are briefly treat-
ed of several subjects relating to
natural philosophy in an experimental way... [24], 158, 28p.
London, 1691.
W203.   UP

81. ----. Experiments and notes
    about the mechanical origine
or production of electricity. 38p.
London, 1675.
Mp131; (R); W178.   APS

82. ----. Experiments and notes
    about the mechanical produc-
tion of magnetism. 20p. London,
1676.
W178   APS

83. ----. Tractatus, in quibus
    continentur suspiciones de
latentibus quibusdam qualitatibus
aeris; una cum appendice de magne-
tibus coelestibus, nonnullisque
argumentis aliis,... [4] 87p.
Geneva, 1680.
Mp132.   APS

84. ----. Works...to which is
    prefixed the life of the
author [by Thomas Birch]. New ed.
6v. plates., port. London, 1772.
(W188a).   APS

85. [BRAGADIN, FRANCESCO MARIA].
    Dubbi sull'efficacia de con-
duttori elettrici [Four letters
from Bragadin in answer to one by
Giacomo Scaguller upon an accident
by lightning to the Palazzo Gritti
at Visnadel]. 122p. 1pl. Venice,
1795.
Ms. notes.
Three letters of G. Toaldo: pp.112-
122.
R.   APS

86. ----. Risposta dell'autore
    dei Dubbi sull'efficacia dei
conduttori, all giunta al giornale
Astro-meterologico del pubblico
Prof. Giuseppe Toaldo. 31p. n.p.,
[1796?].

No t.p.
Ms. notes.
R.   APS

87. BRESSY, JOSEPH, 1758-1838.
     Du grandinisme, pour ré-
pondre à la question de l'Academie
royale des sciences, sur la fièvre
continue. En deux parties; la
première traité de la fièvre continue, et la seconde de l'électricité vitale. viii, 80p. Paris,
1835.
Not in MRW.   CP

88. ----. Essai sur l'électri-
     cité de l'eau. viii, 178p.
2pl. Paris, [1797].
Mp324, 557; R.   APS

89. [BRISSON, MATHURIN JACQUES],
     1723-1806. Rapport fait a
l'Académie Royale des Sciences, sur
la machine électrique nouvellement
inventée par M. Walckiers de St.
Amand. 29p. 1 colored plate.
[Paris, 1784].
Extrait des registres de l'Académie
Royale des Sciences...
Not in MRW.   APS

90. ----. Traité élémentaire, ou
     Principes de physique...3rd
ed. 3v. plates. Paris, 1800.
Presented by W. P. C. Barton.
Also 4th ed. Paris, 1803.
Not in MRW.   APS

91. BROOK, ABRAHAM, fl. 1789.
     Miscellaneous experiments and
remarks on electricity, the air
pump, and the barometer: with the
description of an electrometer of
a new construction. xiii, [2],
211, [3]p. 3pl. Norwich, 1789.
Mp231; R; W553.   LCP

92. BROWNE, SIR THOMAS, 1605-1682.
     Pseudodoxia epidemica: or, En-
quiries into very many received
tenets, and commonly presumed
truths. [16], 386p. London,
1646.
W123.   CP, LCP

93. BRUNO,   DE
     Recherches sur la direction du
fluide magnétique, dédiées à Monsieur, frère du roi. viii, 206p.
8pl. Amsterdam & Paris, 1785.
Mp556; R; W527.   APS

BRYDONE, PATRICK: see 499

94. CABEO, NICCOLO, 1585-1650.
     Philosophia magnetica in qua
magnetis natura penitus explicatur, et omnium quae hoc lapide cer-
nuntur, causae propriae asseruntur.
[9], 412p. illus., front. Coloniae,
1629.
Mp110, R; W97.   FI

95. CALANDRELLI, GIUSEPPE, 1749-
     1827. Ragionamento sopra il
conduttore elettrico quirinale...
36p. Rome, 1789.
(W554).   APS

CALDWELL, CHARLES: see 62

96. CANTON, JOHN, 1718-1772.
     Electrical experiments, with
an attempt to account for their
several phenomena; together with
some observations on thunder-clouds
...from the Philosophical transactions. pp.143-154. [London],
1753.
With: Franklin B. Experiments &
observations on electricity...
1751.
Mp205; W367a.   APS

see also: 178, 366

CAULLET DE VEAUMOREL: see 383

97. CAVALLO, TIBERIUS, 1749-1809.
     A complete treatise of elec-
tricity in theory and practice,
with original experiments. xvi,
viii, 412p. 3pl. London, 1777.
Page 404 erroneously numbered 440.
Mp243 etc; R; W463.   LCP, APS

98. ----. Complete treatise on
     electricity,... 2d ed. xxiv,
495, [8]p. 4pl. London, 1782.
Mp243-5 etc.   FI, UP, APS

10

99. ----. A complete treatise of
electricity... 3rd ed. 2v.
plates. London, 1786.
Not in MRW.   FI

100. ----. A complete treatise on
electricity,... 4th ed. Containing the practice of medical
electricity, besides other additions & alterations...volume 3, entirely new, containing the discoveries and improvements made since
the third edition. 3v. 6pl.
London, 1795.
Presented by John Vaughan 1840.
Mp243; R; W463a.   LCP, APS

101. ----. The elements of natural or experimental philosophy. 4v. plates. London, 1803.
Autograph: Robert Maskell Patterson, 1812 (APS).
R; W648.   LCP, APS

102. ----. Elements of natural or
experimental philosophy. 1st
Am. ed. with additional notes...
2v. plates. Philadelphia, 1813.
W648a.   LCP

103. ----. The elements of natural or experimental philosophy. 4th Am. ed., with additional
notes,...by F. X. Brosius. 2v. in
1. 783p. plates. Philadelphia,
1829.
Autograph: Elihu Thomson
Not in MRW.   FI

104. ----. Essay on the theory
and practice of medical electricity. xvi, 112p. 1pl. London,
1780.
Mp243; R; W489.   CP

105. ----. An essay on the theory
and practice of medical electricity; designed to render it more
useful, as well as more easy, both
to patients and practitioners; and
published for the information of
those who know only the old method.
To which are added, Some authentic
cases and new experiments... xxiii,

155p. 1pl. London & Dublin, 1781.
(R).   APS

106. ----. Traité complet d'électricité. Tr. de l'Anglois
[par l'Abbé de Silvestre] sur la
seconde & dernière edition de l'auteur, enrichie de ses nouvelles expériences. xxiv, 343p. 4pl. Paris,
1785.
R.   FI, LCP, APS

107. ----. Trattato completo
d'elettricità teorica e
pratica, con sperimenti originali,
del Signore Tiberio Cavallo, in
italiano dall'originale inglese.
Con addizioni e cangiamenti fatti
dall'autore. xx, 511, [1]p. 3pl.
Florence, 1779.
R.   LCP, APS

108. ----. A treatise on magnetism, in theory and practice,
with original experiments. 2d ed.,
with a supplement. xiv, 343, [9],
[4], 72p. 6pl. London, 1795.
Mp244; (R); (W540).   FI, APS

109. ----. Treatise on magnetism
... 3rd ed., with a supplement. xi, 335p. 3pl. London,
1800.
Mp584; R; W540a.   LCP, CP

110. ----. Vollständige Abhandlung der theoretischen und
praktischen Lehre von der Elektricität... 284p. plates. Leipzig,
1779.
R.   FI

111. ----. Vollständige Abhandlung...aus dem Englischen
übersetzt [by J. S. T. Gehler].
2d ed. [14], 328, [10]p. 4pl.
Leipzig, 1783.
R.   LCP, APS

112. ----. Vollständige Abhandlung...nebst eignen Versuchen
...aus dem Englischen übersetzt.
[6], 255, [11]p. 3pl. Graez, n.d.
Not in MRW.   LCP

CAZELES: see MASARS DE CAZELES

113. CEPPI, LUIGI ANTONIO.
Dissertazione serio-giocosa sull'elettricità artifiziale, nella quale se sottomette alla giudiziosa critica degli studioso filoelettri dell'uno, e dell'altro emisfero. Una nuova teoria fisico-meccanica... 345, [3]p. 13pl. Vercelli, 1784.
Ms. notes (APS).
R.   LCP, APS

114. CHERNAK, LADISLAUS.
Dissertatio physica de theoria electricitatis Franklini, in explicando experimento Leidensi, cum dubiis eandem non mediocriter infirmantibus. [2], 46, [1]p. Groningen, 1771.
Thesis read at Haarlem, May 4, 1771.
R.   APS

115. CHIARE ISTRUZIONI per construire ed innalzare siguri conduttori. Tr. dall'Inglese. x, 53p. Naples, 1794.
Not in MRW.   APS

116. CHIGI, ALESSANDRO.
Dell' elettricità terrestra-atmosferica... iv, 170p. Siena, 1777.
Mp585; R.   LCP, APS

117. ----. Lettera ad un amico... sopra il fulmine caduto nel dì 18. Aprile del corrente anno 1777. nella spranga posta nella torre del Palazzo pubblico della città di Siena. Che può servire di supplemento alla Dissertazione dell' elettricità terrestre-atmosferica... 16p. Siena, 1777.
R; W464.   LCP

118. CHURCHMAN, JOHN, 1753-1805.
An explanation of the magnetic atlas, or variation chart, hereunto annexed; projected on a plan entirely new, by which the magnetic variation on any part of the globe may be precisely determined, for any time, past, present, or future: and the variation and latitude being accurately known, the longitude is of consequence truly determined. x, [3], 14-46, 5p. 2 tables. Philadelphia, 1790.
John Dickinson's copy (HSP).
Mp315; R; W562.   FI, LCP, HSP, APS

119. ----. Magnetic atlas; or, Variation charts of the whole terraqueous globe... vii [1] 80p. 1pl. 2 charts. London, 1794.
Mp315; R; W588.   FI, LCP, APS

120. ----. The magnetic atlas... 4th ed. xviii, 86p. plates. London, 1804.
Author's copy, with portraits, clippings & ms. letters bound in.
Mp586; R; W588a.   LCP

see also: 1283

121. CIGNA, GIOVANNI FRANCESCO, 1734-1790. De novis quibusdam experimentis electricis. [31]-72p. [Turin, 1766].
From Miscellanea taurinensia...de la Soc. R. de Turin, v.3.
Benjamin Franklin's copy.
Mp224; R.   HSP

COLDEN, DAVID: see 178

122. [CONSTANTINI, GIUSEPPE?].
Difesa della comune, ed antica sentenza che i fulmini discendano dalle nuvole contro l'opinione del Sig. Marchese Scipione Maffei che si formino al basso, ed ascendano. Riflessioni dell'autore delle Lettere critiche appoggiate alla ragione, ed alla sperienza, con un discorso in fine intorno alla somiglianza della forza del fuoco de' fulmini, e della luce elettrica. xii, 184p. Venice, 1749.
R.   LCP

123. COOPER, M   .
A philosophical enquiry into

12

the properties of electricity...in
which is contain'd a confutation
of the solutions which have been
hitherto given of it, and the most
probable reason of the late surprizing experiments. In a letter
to a friend. 32p. London, 1746.
Mp746; W337.   LCP

124. CORNACCHINI, PIETRO.
       Della pazzia; dissertazione
e due discorsi accademici sopra la
medicina elettrica con alcune cure
fatte per mezzo della medesima.
[5], 159p. Siena, 1758.
Ms. notes.
Not in MRW.   CP

COSNIER: see LEDRU

125. CREVE, JOHANN CASPAR IGNAZ
       ANTON, 1769-1853. Beiträge
zu Galvanis Versuche über die
Kräfte der thierischen Elektrizität auf die Bewegung der Muskeln.
104p. Frankfurt & Leipzig, 1793.
R.   APS

126. CROKER, TEMPLE HENRY, 1730?-
       1790? Experimental magne-
tism, or, The truth of Mr. Mason's
discoveries...that there can be no
such thing in nature, as an in-
ternal central loadstone, proved
and ascertained... x, 72p. plate,
front. London, 1761.
Joseph Priestley's copy.
R; W404.   LCP

127. CUYPERS, C   .
       Verslag van zekere behande-
ling waar door glaze schyven voor
electrizeer machines. 37p. The
Hague, 1778.
(W471).   FI

DALIBARD, THOMAS FRANÇOIS: see
       171-2

128. DALTON, JOHN, 1766-1844.
       Meteorological observations
and essays. xvi, 208p. London,
1793.

Presented by the author, Nov. 1802.
Mp307; W582.   APS

129. ----. Meteorological obser-
       vations and essays... 2d ed.
xx, 244p. Manchester, 1834.
Presented by Literary & philosophical society of Manchester.
(Mp307); R; W582a.   APS

130. [DARWIN, ERASMUS], 1731-1802.
       The botanic garden; a poem,
in two parts...with philosophical
notes. v.p. plates, fronts.
London, 1791.
Part II is 2d ed.; separate t.p.
dated 1790.
APS has several other editions.
W555.   APS

131. ----. Phytologia: or, The
       philosophy of agriculture
and gardening... viii, 556, [11]p.
12pl. Dublin, 1800.
W621.   APS

132. ----. The temple of nature;
       or, The origin of society:
a poem, with philosophical notes.
256p. New York, 1804.
Not in MRW.   APS

DAVY, SIR HUMPHRY: see 393.

133. DE FREMERY, NICOLAUS CORNE-
       LIUS, 1770-1844. ...De ful-
mine... [2], 96, [2]p. 1pl.
Leyden, 1790.
Inaugural dissertation.
Mp556; R.   APS

134. DESAGULIERS, JEAN THÉOPHILE,
       1683-1744. A dissertation
concerning electricity... 6, [1],
50p. London, 1742.
Mp167; R; W306.   LCP

135. ----. Dissertation sur l'élec-
       tricité des corps... 28,
[2]p. Bordeaux, 1742.
Mp167; R.   APS

DESMAREST, NICOLAS: see 244

136. DÍAZ DE GAMARRA Y DÁVALOS,
JUAN BENITO, 1745-1783.
Elementa recentioris philosophiae
...Exmelioris notae recentioribus
philosophis excerptum, congestum,
adornatum ad usum scholaris juven-
tutis in perillustri Collegio sale-
siano apud pp. presbyteros seculares
Congregationis oratorii s. Philippi
Nerii michaelopoli in Nov. Hisp...
2v. plates, port. Mexico, 1774.
Not in MRW.   APS

137. DIGBY, SIR KENELM, 1603-1665.
Of bodies, and of mans soul.
To discover the immortality of
reasonable souls. With two discourse Of the powder of sympathy,
and Of the vegetation of plants.
2v. in 1. illus. London, 1669.
V.2 has title: Second treatise:
declaring the nature and operations of mans soul...
(Mp121); (W156).   APS

138. DISSERTAZIONE FILOSOFICA sopra la folgore, e suono delle
canpane, colla quale si prova che
il suono delle medesime nulla in-
fluisce allo scoppiare, nè ad at-
trarre a sè la folgore. 29p.
[Lucca, 1787].
Signed: G.B.B.R. (authority not
known: Giovanni Battista Botti
Rettore di Maranello).
Not in MRW.   APS

139. DIVISCH, PROCOPIUS, 1696-1765.
Langst verlangte Theorie von
der meteorologischen Elektricität
welche er selbt Magiam naturalem
benahmet; samt einem Anhang von
Gebrauch der elektrischen Gründe
zur Chemie. 180p. Frankfurt,
1768.
R.   FI

DOLLOND, PETER: see 687

140. DOMIN, JOSEPH FRANCIS, 1754-
1819. Lampadis electricae
optimae notae descriptio eaque
utendi ratio... 22p. 1pl. Pest,
1799.
R.   APS

141. DONNDORFF, JOHANN AUGUST,
1754-1837. Die Lehre von
der Elektricität, theoretisch und
praktisch auseinander gesezt, zum
gemeinnützigen Gebrauch, auch für
solche, die keine Gelehrte sind.
2v. xviii, [4], 977p. 7pl. Erfurt,
1784.
R.   APS

142. ----. Sendschreiben an...
den Herrn Grafen von Borcke
...ueber einige Gegenstände der
Elektricität. 64p. 1pl. Quedlinburg, 1781.
R.   LCP

143. DUFAY, CHARLES FRANÇOIS DE
CISTERNAY, 1698-1739.
...Anmerckungen über verschiedene
mit dem Magnet angestellete Versuche...wie auch des Herrn von
Reaumur Versuche, womit er beweiset, dass der Stahl und das Eisen
leichtlich magnetisch werden, wenn
man sie gleich mit Keinem Magnet
bestrichen, beygefüget worden Aus
dem Frantzösischen... 196, [26]p.
plate. Erfurt, 1748.
R.   APS

144. DUHAMEL, JEAN BAPTISTE, 1624-
1706. De corporum affec-
tionibus cum manifestis, tum oc-
cultis, libri duo: seu promotae
per experimenta philosophiae spe-
cimen,... 2v. in 1. [10], 556,
[16]p. Paris, 1670.
R.   APS

see also: 366

145. DUTOUR, ÉTIENNE FRANÇOIS, 1711-
1784. Recherches sur les différens mouvemens de la matiere électrique, dédiées à M. l'Abbé Nollet...
xxiii, 318 [6]p. 4pl. Paris, 1760.
Presented by Bern Dibner, August
1967.
Mp170; R.   APS

146. [EANDI, GIUSEPPE ANTONIO
FRANCESCO GIROLAMO], 1735-1799.
Memorie istoriche intorno gli
studi del padre Giambatista Beccaria
... 161p. [Turin], 1783.

14

Includes bibliography of Beccaria.
Mp294; R.   LCP, APS

147. EBERHARD, CHRISTOPH, 1655-
      1730. Specimen theoriae
magneticae, quo ex certis principiis
magneticis ostenditur vera et uni-
versalis methodus inveniendi lon-
gitudinem et latitudinem... 13,
[1] 50p. 1pl. Leipzig, 1720.
Latin and German Text.
R.   APS

148. EHRMANN, FRIEDRICH LUDWIG,
      1741-1800. Beschreibung und
Gebrauch einiger elektrischer Lam-
pen. Aus dem Französischen von
ihm selbst übersetzt. Mit Anmerkungen und einem Anhange. 36p.
1pl. Strassburg, 1780.
Not in MRW.   APS

149. EPP, FRANZ XAVIER, 1733-
      1789. Aghandlung von dem
Magnetismus der natürlichen
Electricität. [16], 128p. 2pl.
Munich, 1777.
R.   APS

150. ERXLEBEN, JOHANN CHRISTIAN
      POLYKARP, 1744-1777. An-
fangsgründe der Naturlehre;...
4. Aufl, mit Zusätzen von G. C.
Lichtenberg. 1vi, 710, [38]p.
plates. Göttingen, 1787.
Not in MRW.   APS

151. ESSER, FERDINAND.
      Abhandlung über die Sicher-
heit und Einrichtung der Blitzableiter. [5], 102p. Münster,
1784.
R.   APS

152. ÉTIENNE,   D'
      Recherche pour augmenter la
force de l'électricité, de toutes
sortes de machines, par le moyen
d'une armure adaptée au premier
conducteur. 8p. [Paris, 1775].
From Journal de physique, v.6.
Benjamin Franklin's copy.
Not in MRW.   APS

153. [----.] Recherches pour

améliorer les machines élec-
triques. Par l'auteur de l'armure
et des isolamens alternatifs. 7p.
[Paris], n.d.
From Journal de physique.
Benjamin Franklin's copy.
Not in MRW.   APS

154. ----. Recherches sur le
      moyen de perfectionner les
isolamens pour toutes sortes de
machines électriques. 9p. [Paris,
1775].
From Journal de physique, Oct. 1775.
Benjamin Franklin's copy.
Not in MRW.   APS

155. EULER, JOHANN ALBRECHT, 1734-
      1800. Disquisitio de causa
physica electricitatis ab Academie
scientiarum imperiali Petropolitana
praemio coronata... Una cum aliis
duabus dissertationibus de eodem
argumento. [by Beraud & Frisi,
q.v.] 144p. 2pl. St. Petersburg,
[1755].
Benjamin Franklin's copy.
R; W386.   HSP

156. ----. Dissertationes selec-
      tae; Jo. Alberti Euleri,
Paulli Frisii, et Laurentii Beraud,
quae ad Imperialem Scientiarum Pet-
ropolitanam Academiam an. 1755 mis-
sae sunt, cum electricitatis caussa
& theoria praemio proposito, quaere-
retur. 15, 204p. 1pl. St. Petersburg & Lucca. 1757.
R.   APS

157. EULER, LEONHARD, 1707-1783.
      Letters on different sub-
jects in physics and philosophy.
Tr. from the French by Henry Hunt-
er. 2v. plates. London, 1795.
Also 2d ed. 1802.
R; W635.   LCP

158. FAULWETTER, CARL ALEXANDER,
      1745-1801. Kurze Grundsätze
der Elektricitäts-Lehre. 3v.
Nürnberg, 1790-92.
(R).   FI

159. FAURE, GIOVANNI BATTISTA.

Congetture fisiche intorno alle cagioni fenomeni osservatio in Roma nella macchina elettrica all' illustrissimo signor Giambattista Collicola... xii, 140p. Rome, 1747.
Mp555; R; W339.   LCP, APS

160. FELBIGER, JOHANN IGNATZ VON, 1724-1788. Die kunst Thürme oder andere Gebäude vor den schadlichen Wirkungen des Blitzes durch Ableitungen zu bewahren, angebracht an dem Thurm der saganischen Stiftsund Pfarrkirche. 110, [2]p. 1pl. Breslau, 1771.
R.   LCP

161. FELLER, CHRISTIAN GOTTHOLD. De therapia per electrum. [6], 24p. Leipzig, 1785. Inaugural dissertation.
R.   CP

162. FERGUSON, JAMES, 1710-1776. An introduction to electricity. In six sections... [2], 140p. 3pl. London, 1770.
Mp601; R; W429.   FI, LCP, APS

163. ----. An introduction to electricity... 3rd ed. [2] 140p. 3pl. London, 1778.
Mp601; R; W429b.   LCP, CP

164. FLORENCE. ACCADEMIA DEL CIMENTO. Tentamina experimentorum naturalium captorum in Accademia del cimento...et ab ejus academiae secretario conscriptorum; ex italico in latinum sermonem conversa quibus commentarios, nova experimenta et orationem de methodo instituendi experimenta physica addidit Petrus van Musschenbroek... [by Lorenzo Magalotti]. v.p. 32pl. Vienna, 1756.
(Mp175); R; (W276).   APS

165. FLUDD, ROBERT, 1574-1637. Philosophia Moysaica. In qua sapientia & scientia creationis & creaturarum sacra veréque Christiana...ad admussim et

enucleate explicatur. [5], 152 𝓵 illus. Gouda, 1638.
Mp554; W112.   CP

166. IL FLUIDO elettrico applicato a spiegare i fenomeni della natura. 64p. illus. Ancona, 1772.
Not in MRW.   APS

167. FOLLINI, GIORGIO, 1751?-1831. Teoria elettrica brevemente esposta ad uso della studiosa gioventù. 164, [4]p. 2pl. Ivrea, 1791.
Pp. 25-40 lacking (APS).
R.   FI, APS

168. FONDA, GIROLAMO MARIA. Sopra la maniera di preservare gli edifici dal fulmine memoria fisica...in occasione del fulmine caduto sopra la cupola della chiesa contigua,... 30p. front. Rome, 1770.
Mp253; R.   LCP, APS

169. FOWLER, RICHARD, 1765-1863. Experiments & observations relative to the influence lately discovered by Galvani & commonly called animal electricity. iii, 176p. Edinburgh, 1793.
Mp306; R; W583.   FI, CP

170. FRANKLIN, BENJAMIN, 1706-1790. ...Briefe von der Elektricität. Aus dem Engländischen übersetzet, nebst Anmerckungen von J. C. Wilcke. [26], 354p. front. Leipzig, 1758.
Mp216; R; W367f.   FI, UP, APS

171. ----. Erweitertes Lehrgebäude der natürlichen Elektrizität. Für jedermann fasslich und deutlich dargestellt durch D.L.G. 100p. Vienna, 1790.
R.   APS

172. ----. Expériences et observations sur l'électricité faites à Philadelphie en Amèrique ...& communiquées dans plusieurs

16

lettres à M.P. Collinson de la Société royale de Londres. Tr. de
l'anglois. 24, lxx, [9], 222,
[29]p. 1pl. Paris, 1752.
Inscription: Thos. Hewson Bache
from Duane Williams (APS).
"Histoire abregée de l'électricité," par Thomas François Dalibard, pp. i-lxx.
R.   FI, APS, HSP

173. ____. Expériences et obser-
           vations...2. éd...augmentée
d'un supplément considérable...
avec des notes et des expériences
nouvelles. Par M. d'Alibard. 2v.
2pl. Paris, 1756.
R.   FI, APS

174. ----. Experiments & obser-
           vations on electricity, made
at Philadelphia in America...com-
municated in several letters to
Mr. P. Collinson of London. [4],
86p. 1pl. London, 1751.
Mp193ff; R.   FI, UP, HSP, APS

175. ----. Experiments and obser-
           vations on electricity,...
to which are added letters and
papers on philosophical subjects
...4th ed. 496, [16]p. illus,
7pl. London, 1769.
Ms. notes (c.2, APS)
Mp193ff; R.   FI, UP, LCP, HSP,
                CP, APS

176. ----. Experiments and ob-
           servations...to which are
added letters and papers on
philosophical subjects...5th ed.
v, 514, [16]p. 7pl, front.
London, 1774.
Benjamin Franklin's copy (APS)
R; W367c.   FI, LCP, APS

177. ----. New experiments and
           observations on electricity,
made at Philadelphia in America &
communicated in several letters
to Peter Collinson of London.
154p. illus, 1pl. London, 1754.
Pts. 1 & 2 are 2d ed., pt. 3 is
1st ed.
Mp193ff; R.   FI, HSP, APS

178. ----. New experiments and
           observations...Communicated
to P. Collinson...And read at the
Royal society June 27, and July 4,
1754. To which are added a paper
on the same subject by J. Canton...
read at the Royal society Dec. 6,
1753; and another in defence of Mr.
Franklin against the Abbé Nollet,
by Mr. D. Colden of New York.
Part III. [109]-154p. illus.
London, 1754.
Bound with his Experiments and
observations...1751.
Mp193ff; R.   UP, APS

179. ----. New experiments and
           observations...3rd ed. Parts
I & II. 1pl. London, 1760-62.
R.   UP, APS

180. ----. New experiments and
           observations... 3 parts.
iv, 154p. illus., 1pl. London,
1760-65.
Parts I & II are 3rd ed. Part III
is 4th ed.
R.   APS

181. ----. Oeuvres de M. Franklin
           ...Tr. de l'anglois sur la
quatrième édition. Par M. Barbeu
Dubourg. Avec des additions nouvelles. 2v. plates, port. Paris,
1773.
Bookplate: Philip Tidyman, presented by him, 1846.
R.   APS

182. ----. Opere filosofiche di
           Beniamino Franklin, nuova-
mente raccolte, e dall'originale
inglese recate in lingua italiana.
125p. 3pl. Padua, 1783.
R.   LCP, APS

183. ----. Scelta di lettere e
           di opuscoli del Signor Ben-
iamino Franklin, tr. dall'inglese.
99p. Milan, 1774.
Not in MRW.   LCP, APS

184. ----. Supplemental experi-
           ments and observations on
electricity, part II. Made at

Philadelphia in America,...and communicated in several letters to P. Collinson. [87]-107, [1]p. London, 1753.
Bound with his Experiments and observations...1751 (c.2, APS). (367a).   HSP, APS

see also: 519, 1059-61, 1064-67, 1070-72, 1249-50, 1286.

185. FREKE, JOHN, 1688-1756.
     Essai sur la cause de l'electricité, où l'on examine pourquoi certaines choses ne peuvent pas être électrifiées. Et quelle est l'influence de l'électricité dans ...le corps humain...Adressé en forme de lettre à M. Guill. Watson ...tr. de l'anglois. 2. éd. avec un supplément. viii, 52p. 2pl. Paris, 1748.
In Recueil de traités sur l'électricité, v.3.
Not in MRW.   UP, LCP, APS

186. ----. An essay to shew the cause of electricity; and why some things are non-electricable. In which is also consider'd its influence in the blasts on human bodies, in the blights on trees, in the damps in mines; and as it may affect the sensitive plant, &c. In a letter to Mr. William Watson, F.R.S. viii, 51p. London, 1746.
Mp201.   LCP

187. ----. An essay to shew the cause of electricity...2d ed. with an appendix. viii, 64p. London, 1746.
Mp201; R; W325.   LCP, APS

188. ----. An essay to shew the cause of electricity...3rd ed. with an appendix. 196p. London, 1752.
Not in MRW.   FI

189. ----. Treatise on the nature and property of fire. In three essays. I. Shewing the cause of vitality... II. On electricity. III. Shewing the mechanical cause

of magnetism; and why the compass varics in the manner it does.
viii, 196p. London, 1752.
Mp201; R; W371.   LCP

190. FRISI, PAOLO, 1728-1784.
     De caussa electricitatis dissertatio. 41-131p. [St. Petersburg & Lucca, 1757].
In J. A. Euler's Dissertationes selectae...1757 (q.v.)
R.   APS

191. ----. De existentia et motu aetheris seu de theoria electricitatis ignis et lucis, dissertatio. pp. 29-94. [St. Petersburg, 1755].
With J. A. Euler's Disquisitio... (q.v.)
Benjamin Franklin's copy.
W387.   HSP

192. ----. Opuscoli filosofici. [4,4], 118p. Milan, 1781.
Contains his "Dei conduttori elettrici."
R.   APS

193. GALISTEO, JUAN.
     Remedio natural para precaverse de los rayos y de sus funestos efectos. Secreto tan util, como curioso, sacado de las repetidas observaciones, y experiencas, que sobre la analogia de la electricidad con la materia de los rayos...
7p. Mexico, [1757].
Not in MRW.   CP

194. GALITZIN, DMITRI ALEXIEVITCH, PRINCE, 1738-1803. Sendschreiben an die kaiserliche Akademie der Wissenschaften zu St. Petersburg über einige Gegenstände der Electricität. 56p. 3pl. Münster, 1780.
R; W491.   APS

195. GALVANI, LUIGI, 1737-1798.
     De viribus electricitatis in motu musculari commentarius cum Joannis Aldini dissertatione et notis. Accesserunt epistolae ad animalis electricitatis theoriam pertinentes. xxvi, 80p. 3pl.

Modena, 1792.
Mp283; R; W570a.   CP

196. [----]. Dell'uso e dell'at-
      tività dell'arco conduttore
nelle contrazioni dei muscoli.
168p. Bologna, 1794.
R.   APS

197. ----. Memorie sulla elet-
      tricità animale...al celebre
abate Lazzaro Spallanzani...ag-
giunte alcune elettriche esperienze
di Gio. Aldini... 105, [5]p. 2pl.
Bologna, 1797.
R; W606.   APS

198. ----. Opere edite ed inedite
      ...raccolte e pubblicate per
cura dell'Accademia delle scienze
dell'istituto di Bologna. 120,
505, 58p. 9pl. port. Bologna,
1841-42.
R; W999.   FI, CP, APS

199. [----]. Supplemento al
      trattato dell'uso e dell'
attivitá dell'arco conduttore
nelle contrazioni de'muscoli. 23,
[1]p. n.p., [1794].
R.   APS

see also: 589, 740

200. GARDINI, GIUSEPPE FRANCESCO,
      1740-1816. L'applicazione
delle nuove scoperte del fluido
elettrico agli usi della ragione-
vole medicina. Dissertazione
scritta...al dottore Carlo Gan-
dini. xiv, 248p. Genoa, 1774.
Mp385; R.   LCP

201. ----. De effectibus elec-
      tricitatis in homine. Dis-
sertatio... 141p. Genoa, 1780.
R.   CP

202. GILBERT, WILLIAM, 1540-1603.
      ...De magnete, magneticisque
corporibus, et de magno magnete
tellure; physiologia nova, plu-
rimis & argumentis, & experimentis demonstrata. [8], 240p.
illus. London, P. Short, 1600.

Mp82; R; W72.   FI

203. ----. Tractatus, sive phys-
      iologia nova, De magnete,
magneticisque corporibus et magno
magnete, tellure sex libris...Om-
nia nunc diligenter recognita...opera
& studio Wolfgangi Lochmans. [4,6],
232, [17]p. illus. 12pl. Stettin,
1628.
Mp607, etc.; R; W72a.   APS

204. ----. Tractatus; sive phys-
      iologia nova de magnete,...
[10], 232, [17]p. illus., plates.
Stettin, 1633.
R; W72a.   UP, CP

205. GLANVILL, JOSEPH, 1636-1680.
      Scepsis scientifica: or,
Confest ignorance, the way to science; in an essay of the vanity of
dogmatizing, and confident opinion
... [28], 184p. London, 1665.
Mp127; R; W147a.   APS

206. ----. Scire tuum nihil est:
      or, The authors defence of the
vanity of dogmatizing: Against the
exceptions of the learned Tho.
Albius in his late Sciri. [12],
92p. London, 1665.
W147b.   APS

207. GNILIUS, JOHANNIS HENRICUS.
      Momenta quaedam ex capite phys-
ices experimentalis de magnete sub
praesidio dn. Johannis Philippi
Grauel... 16p. Strasbourg, [1761].
Not in MRW.   APS

208. [GOCLENIUS, RUDOLPH], 1547-
      1628. Trinium magicum, sive
secretorum magicorum opus... [12],
498, [2]p. Frankfurt, 1630.
W100.   CP

GONFIGLIACCHI, PIETRO: see 537

209. GORDON, ANDREW, 1712-1751.
      Phaenomena electricitatis ex-
posita. 88p. 1pl. Erfurt, 1744.
Mp168; R.   LCP

210. ----. Versuch einer Erklärung

der Electricität. [16], 88p. 2pl. Erfurt, [1745].
R; W317.   APS

211. [GRAHAM, JAMES], 1745-1794.
Dr. Graham is now preparing the largest and most elegant medico-electrical-aërial and magnetic apparatus in the world... [16]p.
[Newcastle, 1779].
Benjamin Franklin's copy.
Not in MRW.   HSP

212. ----. The general state of medical and chirurgical practice, ancient and modern, exhibited;...to which are added a great number of recent and remarkable cases and cures, never before published. [6], vi, 152, 33p. Bath, 1778.
Not in MRW.   LCP

213. ----. The general state of medical and chirurgical practice...and to the whole are added, near an hundred recent and remarkable cases, cured... 6th ed. 248p. London, 1779.
Not in MRW.   LCP

214. ----. The guardian goddess of health; or, The whole art of preventing and curing diseases ...to which is added, an account of the...three great medicines prepared at the Temple of health, Adelphi, and at the Temple of hymen, Pall-Mall, London. 38, [2]p.
London, n.d.
Not in MRW.   LCP

215. ----. A lecture on the generation, increase, and improvement of the human species!... with a glowing, brilliant, and supremely delightful description of the structure, and most irresistibly genial influences of the celebrated celestial bed!!!... [4], 22p. London, [1783].
Ms. notes.
Not in MRW.   LCP

216. GRANDAMI, JACQUES, 1588-1672.
Nova demonstratio immobilitatis terrae petita ex virtute magnetica. Et quaedam alia ad effectus & leges magneticas. usumque longitudinum & universam geographiam spectantia, de novo inventa. [8], 170p. illus., plates, front.
La Flèche, 1645.
Mp120; R; W122.   APS

217. GREGORY, GEORGE, 1754-1808.
The economy of nature explained and illustrated on the principles of modern philosophy.
3v. 46pl. London, 1796.
Mp323; W598.   APS

218. GRISELINI, FRANCESCO.
Lettera...al padre D. Angelo Calogiera intorno l'elettricità e alcuni particolari esperienze della medesima. [37]-81p. 1pl. [Venice, 1747].
Bound with: Pivati, G. F., Lettera ... (q.v.)
R.   LCP

219. GROSS, JOHANN FRIEDRICH, 1732-1795. Elektrische Pausen.
[6], 136p. illus. Leipzig, 1776.
R.   LCP

220. ----. Grundsätze der Blitzableitungskunst geprüft und durch einen merkwürdigen Fall erläutert...herausgegeben von Johann Friedrich Wilhelm Widenmann. [10], 228p. 1pl. Leipzig, 1796.
R.   APS

221. GUDEN, PHILIPP PETER, 1722-1794. Von der Sicherheit wider die Donner-Stralen... [14], 200p. Göttingen & Gotha, 1774.
R.   APS

222. GUERICKE, OTTO VON, 1602-1686.
Experimenta nova (ut vocantur) Magdeburgica de vacuo spatio... variisque aliis experimentis aucta ... [13], 244, [4]p. illus., 2pl.
Amsterdam, 1672.
Mp126; R; W170.   APS

20

223. GÜTLE, JOHANN CONRAD, b.1747.
Kleine Elektrizitätslehre,
oder Beschreibung einiger kleiner
Elektrisirmaschinen und Apparate,
zum Gebrauch für Schulen und Haus-
lehrer und zur Erklärung der Lehre
der künstlichen und Luftelektrizität eingerichtet. xvi, 223p. 4pl.
Nürnberg, 1798.
R.   APS

224. ----. Kunstkabinet verschie-
dener mathematischer und physi-
kalischer Instrumente und anderer
Kunstsachen... 2 parts in lv. 31,
56p. 7pl. Nürnberg, 1792.
Not in MRW.   APS

225. ----. Versuche Unterhaltun-
gen und Beluftigungen aus der
natürlichen Magie zur Lehre zum
Nutzen und zum Vergnügen bestimmt.
[16], 358p. 11pl. Leipzig & Jena,
1791.
Not in MRW.   LCP

226. ----. Verzeichnis eines elektrischen Maschinen und Experimenten Kabinets. 94p. n.p., 1779.
Not in MRW.   APS

227. ----. Verzeichnus [sic]
eines physikalisch und mate-
matischen Maschinen, Instrumenten
und experimenten Kabinets. [47]-
94p. n.p., 1780.
With his Verzeichnis eines elektrischen Maschinen...1779 (q.v.).
Not in MRW.   APS

228. GUYOT, EDMÉE-GILLES, 1706-
1786. Nouvelles récréa-
tions physiques et mathematiques,
contenant toutes celles qui ont été
découvertes & imaginées dans ces
dernier temps, sur l'aiman, les
nombres, l'optique, la chymie, &c
... 4v. col. plates. Paris,
1769-70.
Mp224; W426.   APS

229. HALE, SIR MATTHEW, 1609-
1676. Magnetismus magnus:
or, Metaphysical and divine con-
templations on the magnet, or loadstone. [1], iv, [2], 159p. London, 1695.
Mp554; W212.   UP

230. HALIDAY, WILLIAM.
...De electricitate medica...
48p. Edinburgh, 1786.
Inaugural dissertation.
Inscription by author to Mr. Gib-
bons.
Not in MRW.   APS

231. HAMBERGER, GEORG ERHARD,
1697-1755. Programma inaugu-
rale de partialitate acus magneticae quo ad audiendam orationem in-
auguralem de atmosphaera lunae...
[8]p. Jena, [1727].
R.   APS

232. [HARRINGTON, ROBERT], fl.
1815. A new system on fire
and planetary life; shewing that
the sun and planets are inhabited
...also, an elucidation of the
phaenomena of electricity and magnetism. iv, 75p. London, 1796.
R; W599.   LCP

233. HARTMANN, JOHANN FRIEDRICH,
d. 1800. Abhandlung von der
Verwandschaft und Aehnlichkeit der
elektrischen Kraft mit den er-
schrecklichen Luft-Erscheinungen
entworfen. [40], 253p. 1pl.
Hannover, 1759.
Mp216; R.   APS

234. ----. Die angewandte Electricität bey Krankheiten des
menschlichen Körpers. [28], 304p.
Hannover, 1770.
R.   APS

235. ----. ...Anmerkungen über
die nöthige Achtsamkeit bey
Erforschung der Gewitter-Elektrici-
tät, nebst Beschreibung eines Elec-
tricität-Zeigers... 57p. 2pl.
Hannover, 1764.
R.   APS

236. ----. Encyclopaedie der
elektrischen-Wissenschaften
als eine Vorbereitung zur naeheren
Kenntnis der Elektricitaet. tabellarisch entworfen. 256p.

Bremen, 1784.
R.   APS

237. ----. Die natürliche Luft-
      Elektricität der Atmosphäre.
tabellarisch entworfen... 39p.
Hannover, 1779.
R.   APS

238. HAÜY, RENÉ JUST, 1743-1822.
      An elementary treatise on
natural philosophy. Tr. from the
French...by Olinthus Gregory...
with notes by the translator. 2v.
plates. London, 1807.
R; W684.   LCP, APS

239. ----. Exposition raisonée de
      la théorie de l'électricité
et du magnétisme d'apres les principes de M. Aepinus. xxvii, [5],
235, [3]p. 4pl. Paris, 1787.
Presented by Robert Maskell Patterson, 1817 (APS, c.1).
Mp286; R; W541.   FI, LCP, APS

240. ----. Nouvelles observa-
      tions sur la faculté con-
servatrice de l'électricité acquise
à l'aide du frottement. 7p.
Extrait du Journal de physique,
décembre, 1819.
Inscribed by author.
Not in MRW.   APS

241. ----. Sur l'électricité des
      minéraux. 8p. 1pl.
From Annales du Muséum d'histoire
naturelle, v.XV, 1810.
Inscription by author to Dr. Barton.
(R).   APS

242. ----. Traité élémentaire de
      physique...Ouvrage destiné
pour l'enseignement dans les ly-
cées nationaux. 2v. plates.
Paris, 1803.
Also 2d ed. Paris, 1806 (APS, LCP)
(Mp353); R.   APS

243. HAUKSBEE, FRANCIS, d. 1713?
      Esperienze fisico-mecaniche
sopra vari soggetti contenenti un
racconto di diversi stupendi fenomeni intorno la luce e l'elettri-

cità producibile dallo strofina-
mento de' corpi...Con spiegazioni
di tutte le macchine...Tr. dall'
idioma inglese. [11], 162p. plates.
Florence, 1716.
R.   LCP

244. ----. Experiences physico-
      mechaniques sur différens
sujets, et principalement sur la
lumiere et l'électricité, produites
par le frottement des corps. Tr...
par M. de Brémond...Revûes et mises
au jour, avec un discours prélimi-
naire, des remarques & des notes,
par M. Desmarest. 2v. plates.
Paris, 1754.
R; W232b.   APS

245. ----. Physico-mechanical ex-
      periments on various sub-
jects. Containing an account of several surprizing phenomena touching
light and electricity, producible
on the attrition of bodies...Together
with the explanations of all the
machines... [12], 194p. plates.
London, 1709.
Mp150; R; W232.   LCP, APS

246. ----. Physico-mechanical ex-
      periments...To which is added
a supplement, containing several
new experiments... 2d ed. [14],
336p. plates. London, 1719.
T.p. missing; photograph copy inserted (APS).
Autograph: E. Newton Harvey (APS)
R; W232a.   FI, APS, LCP

247. HAUSEN, CHRISTIAN AUGUST,
      1693-1743. Novi profectus
in historia electricitatis, post
obitum auctoris,...ex msto. ejus
editi. Praemissa est commenta-
tiuncula de vita et scriptis viri
...[by I.C.G.P.]. [3], xii, 49,
[3]p. front. Leipzig, 1743.
Mp168; R; W309.   UP, CP.

248. ----. Novi profectus in
      historia electricitatis,
post obitum auctoris...Accessit V.
C. Henrici de Sanden...dissertatio
De succino, electricorum principe,

22

quam edidit et de vita B. Hausenii,
praefatus est Jo. Christoph.
Gottsched... [16], 128p. front.
Leipzig, 1746.
R; W309a.   APS

249.   HELL, MAXIMILIAN, 1720-1792.
       Anleitung zum nutzlichen
Gebrauch der künstlichen Stahl-
Magneten. 50, [2]p. plate. Vienna,
1762.
R.   APS

250.   HELMONT, JOHANN BAPTIST VAN,
       1577-1644. A ternary of
paradoxes. The magnetick cure of
wounds. Nativity of tartar in
wine. Image of God in man...tr.,
illustrated, and ampliated by Walter Charleton. [46], 144p. London,
1650.
Mp104; R; W130.   LCP, CP

251.   HEMMER, JOHANN JAKOB, 1733-
       1790. Anleitung, Wetter-
leiter an allen Gattungen von Gebäuden auf die sicherste Art an-
zulegen...mit...einem Anhange von
der Verhaltungsregeln zur Gewit-
terzeit...2d ed. xxii, 232p. 1pl.
Mannheim, 1788.
R.   APS

252.   ----. Verhaltungsregeln,
       wenn man sich zur Gewitter-
zeit in keinem bewafneten Gebäude
befindet... 53p. 1pl. Mannheim,
1789.
R.   APS

253.   HENLEY, WILLIAM, d. ca. 1779.
       An account of the death of a
person destroyed by lightning in
the chapel in Tottenham-Court-Road,
and its effects on the building,
as observed by Mr. Wm. Henley, Mr.
Edward Nairne, and Mr. Wm. Jones.
8p. 1pl. London, 1773.
From Phil. Trans.
Benjamin Franklin's copy.
R.   HSP

254.   ----. Experiments concern-
       ing the different efficacy
of pointed and blunted rods, in

securing buildings against the
stroke of lightning... 22p. 1pl.
London, 1774.
Reprinted from Phil. Trans. v.64.
Ms. notes by the author (APS).
Benjamin Franklin's copy (HSP).
R; W2463.   APS, HSP

see also: 436, 466

255.   HERBERT, JOSEPH, EDLER VON,
       1725-1794. Theoria phae-
nomenorum electricorum. [8], viii,
178p. 1pl. Vienna, 1772.
Mp613; R.   LCP

256.   ----. Theoriae phaenomenorum
       electricorum quae seu electri-
citatis ex redundante corpore in
deficiens traiectu, seu sola atmos-
phaerae electricae actione gignun-
tur. 2d ed. 246p. 6pl. Vienna,
1778.
Mp273; R.   LCP, APS

257.   HOADLY, BENJAMIN, 1706-1757.
       Observations on a series of
electrical experiments, By Dr.
Hoadly, and Mr. Wilson. 76p. Lon-
don, 1756.
Autograph: Benjamin Wilson (LCP).
Mp185; R.   LCP, APS

258.   ----. Observations on a
       series of electrical experi-
ments, 2d ed., with alterations...
by B. Wilson. 86p. London, 1759.
W397.   FI

HÖLL: see HELL

259.   HOFFMANN, CARL FERDINAND.
       De morte in fulmine tactis.
40p. Halle Magdeburg, 1766.
Inaugural dissertation.
Not in MRW.   CP

260.   HOOPER, WILLIAM, fl. 1770.
       Rational recreations in
which the principles of numbers
and natural philosophy are clearly
and copiously elucidated, by a
series of easy, entertaining, interesting experiments...4th ed.
4v. col. plates. London, 1794.

V.3 on electricity and magnetism.
(Mp241); (W508).   APS

HOPKINSON, FRANCIS: see 1275

261. HOWARD, EDWARD, OF BERKS.
    Copernicans of all sorts,
convicted: by proving, that the
earth hath no diurnal or annual
motion...To which is annex'd A
treatise of the magnet: as also
how to find the annual variation
of the compass, at land and sea,
mathematically demonstrated, by a
process unknown before, for the
improvement of navigation. [6],
125p. plates. London, 1705.
R; W228.   LCP

HÜBNER, LORENZ: see 507

262. HUMBOLDT, ALEXANDER, FREI-
    HERR VON, 1769-1859. Ex-
périences sur le galvanisme, et en
général sur l'irritation des fibres
musculaires et nerveuses, tr. de
l'allemand, publiée, avec des additions par J.F.N. Jadelot. xlvi,
530p. 8pl. Paris, 1799.
Presented by J. Engles, 1807 (APS).
Mp330; R; W616.   UP, CP, APS

263. ----. Experiencias acerca
    del galvanismo, y en general
sobre la irritacion de las fi-
bras musculares y nerviosas...tr.
del aleman al frances...por J.Fr.
N. Jadelot...y en castellano por
D.A.D.L.M. 2v. in 1. 8pl. Madrid, 1803.
Not in MRW.   APS

264. ----. Kosmos...5v. in 6.
    Stuttgart & Tübingen, 1845-
1862.
Also: Atlas...in 42 Tafeln mit er-
läuterndem Texte. Herausg. von
Traugott Bromme. 2v. Stuttgart,
[1851-54].
R; (W1159).   APS

265. ----. Sur les variations du
    magnétisme terrestre à dif-
férentes latitudes...lu par M.
Biot...[Dec. 17, 1804]. 24p. 3pl.

n.p. [1804].
W663.   APS

266. ----. Über die unterirdi-
    schen Gasarten und die Mittel
ihren Nachteil zu vermindern. Ein
Beytrag zur Physik der praktischen
Bergbaukunde. viii, 346p. Braun-
schweig, 1799.
R.   APS

267. ----. Versuche über die
    gereizte muskel- und nerven-
faser, nebst vermuthungen über
den chemischen process des lebens
in der thier- und pflanzenwelt.
2v. 8pl. Posen & Berlin, 1797.
R.   APS

see also: 660

268.   HUNTER, JOHN, 1728-1792.
    Anatomical observations on
the torpedo... 11p. 1pl. London,
1774.
From Phil. Trans. v.LXIII.
R.   APS

269. IMHOF, MAXIMUS, 1758-1817.
    Theoretisch-praktische An-
weisung zur Anlegung und Erhal-
tung zweckmässiger Blitzableiter.
41p. 3pl. Munich, 1816.
R.   APS

270. ----. Theoria electrici-
    tatis recentioribus experi-
mentis stabilita, unacum positionibus ex universa philosophia
theoretica selectis... 123p.
Haydhusii, 1790.
Mp617; R.   APS

271. INGENHOUSZ, JOHANNES, 1730-
    1799. Anfangsgründe der
Elektricität, hauptsächlich in Beziehung auf den Elektrophor; nebst
einer leichter Art, vermittelst
eines elektrischen Funkens das
Licht anzuzunden, und einem Briefe
in Betref einer neuen entzundbaren
Knauluft, mit Anmerckungen. Tr.
von Niklas Karl Molitor. xvi, 134p.
Vienna, 1781.
R.   LCP

24

272. ----. The Baker Lecture for
the year 1778...[Electrical
experiments to explain how far the
phenomena of the electrophorus may
be accounted for upon the almost
generally received theory of Dr.
Franklin of positive & negative
electricity.] 24p. London, 1779.
Benjamin Franklin's copy.
R.   APS

273. ----. Expériences sur les
végétaux, spécialement sur
la propriété qu'ils possèdent à
un haut degré, soit d'améliorer
l'air quand ils sont au soleil,
soit de le corrompre la nuit,...Tr.
de l'anglois, par l'auteur. Nouvelle éd.... 2v. port. Paris,
1787-89.
Presented by the author.
(R).   APS

274. ----. Lettre...à M. Molitor
...au sujet de l'influence
de l'électricité atmosphérique sur
les végétaux. 16p.
Extrait du Journal de physique,
mai, 1788.
R.   APS

275. ----. Nouvelles expériences
et observations sur divers
objets de physique. 2v. plates.
Paris, 1785-89.
R.   LCP, APS

276. ----. A ready way of light-
ing a candle, by a very mod-
erate electrical spark. 4p.
[London, 1778].
Benjamin Franklin's copy.
R.   APS

277. ----. Vermischte Schriften
phisisch-medizinisch In-
halts. Tr. und herausg. von Niklas Karl Molitor... lxvi, [2],
415, [1]p. plates. Vienna, 1782.
R.   LCP, APS

278. ----. Vermischte Schriften
... 2d ed. 2v. plates,
port. Vienna, 1784.
Presented by the author.
(R).   APS

279. IRVINE, CHRISTOPHER, 1638-
1685. Medicina magnetica:
or, The rare and wonderful art of
curing by sympathy: laid open in
aphorismes; proved in conclusions;
and digested into an easy method
drawn from both... [12], 110p.
[Edinburgh], 1656.
W141.   CP

280. JACQUET DE MALZET, LOUIS
SÉBASTIEN, 1715-1800. Précis de l'électricité; ou, Extrait
expérimental & théoretique des
phénomenes électriques. [5], 235p.
7pl. Vienna, 1775.
R.   APS

JADELOT, JEAN FRANÇOIS NICOLAS:
see 262-3.

281. JALLABERT, JEAN, 1712-1768.
Expériences sur l'électricité, avec quelques conjectures sur
la cause de ses effets. xii, 304p.
4pl. Geneva, 1748.
Presented by Robert Maskell Patterson, 1817 (APS).
Mpl89; R; W349.   FI, LCP, APS

282. ----. Expériences sur l'electricité... xi, [4], 379p.
4pl. Paris, 1749.
R; W349a.   LCP, APS

JONES, JOHN: see 1276

283. JONES, WILLIAM, REV., 1760-
1831. Six letters on electricity. 68p. London, 1800.
R; W622.   LCP

284. KINNERSLEY, EBENEZER, 1711-
1778. A course of experi-
ments, in that curious and entertaining branch of natural philosophy, called electricity; accompanied with explanatory lectures:
in which electricity and lightning, will be proved to be the
same things. 8p. [Philadelphia],
1764.
Not in MRW.   LCP

see also: 1288

285. KIRCHER, ATHANASIUS, 1601-
1680. Magnes siue de arte
magnetica opus tripartitum quo
praeterquam quod universa magnetis...multa hucusque incognita naturae arcana per physica, medica,
chymica et mathematica omnis
generis experimenta recluduntur.
[32], 524, [18], 525-916, [16]p.
illus., 31pl. Rome, 1641.
Ms.notes (APS).
Mp63, 120; R; W116.   LCP, APS

286. -----. ...Magnes, sive De
arte magnetica...2d ed.
... [14], 797, [39]p. illus.,
plates. Cologne, 1643.
Autograph: James Logan (LCP).
Mp621; R; W116a.   FI, LCP, APS

287. -----. Magnes, sive De arte
magnetica...3rd ed. 618,
[14]p. plates. Rome, 1654.
Mp621; R; W116a.   FI, CP

288. -----. Magneticum naturae
regnum, sive disceptatio
physiologica de triplici in natura rerum magnete, juxta tri-
plicem ejusdem naturae gradum di-
gesto inanimato, animato, sen-
sitivo... [16], 201, [7]p. front.
Amsterdam, [1667].
Mp621; (R); W158.   LCP, CP

289. KIRCHHOF, NICOLAUS ANTON
JOHANN, 1725-1800. Be-
schreibung einer Zurüstung welche
die anziehende Krafte der Erde
gegen die Gewitterwolke die Nütz-
lichkeit der Blitzableiter sinn-
lich beweiset... 56p. plate.
Hamburg & Berlin, 1781.
R.   APS

290. KIRCHVOGEL, ANDREAS BERNHARD.
Physikalisch-medicinische
Dissertation von der Wirkung der
Luftelectricität in den mensch-
lichen Körper...aus dem Latei-
nischen übersetzt. [271]-304p.
[Chur & Lindau, 1770].
With F. Bauer's Experimental-
Abhandlung...(q.v.)
R.   APS

291. KLÜGEL, GEORG SIMON, 1739-
1812. Beschreibung der Wirkungen eines heftigen Gewitters,
welches am 12 Jul. 1789 die Stadt
Halle betroffen hat, nebst einer
ausfuehrlichen Erklärung der Ent-
stehung der Gewitter. 64p. Halle,
1789.
R.   APS

292. KNIGHT, GOWIN, 1713-1772.
An attempt to demonstrate,
that all the phaenomena in nature
may be explained by two simple ac-
tive principles, attraction and
repulsion; wherein the attractions
of cohesion, gravity, and magnetism
are shewn to be one and the same;
and the phaenomena of the latter
are more particularly explained.
95p. London, 1748.
R; W350.   UP

293. -----. An attempt to demonstrate... 95p. London,
1754.
R.   LCP

294. -----. Collection of some
papers formerly published in
the Philosophical transactions, relating to the use of Dr. Knight's
magnetical bars, with some notes
and additions. 23p. London, 1758.
R; W394.   LCP, CP

295. KOESTLIN, KARL HEINRICH, 1755-
1783. ...De effectibus electricitatis in quaedam corpora organica... 36p. Tübingen, 1775.
Mp621; W452.   APS

296. KRATZENSTEIN, CHRISTIANUS
GOTTLIEB, 1723-1795. The-
oria electricitatis more geometrico
explicata quam pro gradu magistri
d. Aprilis mdccxxxxvi publice defen-
det... [viii], 62p. front. Halle,
[1746].
Mp170; R; W326.   APS

KRÜNITZ, JOHANN GEORGE: see 440

297. KÜMPEL, JOHANNES ANDREAS
FRIDERICUS. De magnetismo

et minerali et animale. 30p.
Jena, 1788.
Inaugural dissertation.
R.   CP

298. LA BORDE, JEAN BAPTISTE DE,
        d. 1777. Le clavessin élec-
trique; avec une nouvelle théorie
du méchanisme et des phénomenes de
l'électricité. xii, 164p. 2pl.
Paris, 1761.
Mp555; R.   APS

299. LACÉPÈDE, BERNARD GERMAIN
        ÉTIENNE DE LA VILLE SUR
ILLON, COMTE DE, 1756-1825. Essai
sur l'électricité, naturelle et
artificielle. 2v. Paris, 1781.
Autograph: M. Franklin, fils [i.e.
William Temple Franklin] (APS).
Mp273, 556; R; W501.   FI, LCP,
                            APS

LAFOND: see SIGAUD-LAFOND

300. LAMPADIUS, WILHELM AUGUST,
        1772-1842. Versuche und
Beobachtungen über die Elektrizität und Wärme der Atmosphäre, ange-
stellt im Jahre 1792 nebst der
Theorie der Luftelektrizität nach
den Grundsätzen des Hrn. de Luc...
[6], 200p. 3pl. Berlin & Stettin, 1793.
Mp624; (R).   APS

301. LANDRIANI, MARSIGLIO, d.
        1816. Dell'utilità dei con-
duttori elettrici. Dissertazione.
xxxiv, 304p. 1pl. Milan, 1784.
Benjamin Franklin's copy (HSP).
R; W523.   LCP, HSP, APS

302. LANGENBUCHER, JAKOB, d. 1791.
        Beschreibung einer beträcht-
lich verbesserten Elektrisierma-
schine, sammt vielen Versuchen und
einer ganz neuen Lehre vom Laden
der Verstärkung. [26], 268p. 8pl.
Augsburg, 1780.
R.   LCP, APS

303. ----. Richtige Begriffe vom
        Blitz und von Blitzableitern,
aus Erfahrungen gezogen. [6], 44p.

Augsburg, 1783.
Mp389; R.   LCP

304. LA PERRIÈRE DE ROIFFÉ, JACQUES
        CHARLES FRANÇOIS DE, 1694-
1776. Méchanismes de l'électricité
et de l'univers. 2v. plates.
Paris, 1756.
R.   APS

305. LAVOISIER, ANTOINE LAURENT,
        1743-1794. Observations
comuniquées à l'Academie royale
des sciences...sur un effet sin-
gulier du tonnere. 7p. n.p.
[1772?].
No t.p.
Not in MRW.   APS

306. LEDRU, NICOLAS PHILIPPE,
        1731-1807. Rapport de MM.
Cosnier, Maloet, Darcet, Philip,
le Preux, Desessartz, & Paulet...
sur les avantages reconnus de la
nouvelle methode d'administrer
l'électricité dans les maladies
nerveuses, particuliérement dans
l'épilepsie, & dans la catalepsie;
par M. Ledru, connu sous le nom
de Comus... 115p. Paris, 1783.
Mp229, 385; R; W516.   APS

307. LE MONNIER, PIERRE CHARLES,
        1715-1799. Loix du magné-
tisme, comparées aux observations
& aux expériences, dans les différentes parties du globe terrestre,
pour perfectionner la théorie générale de l'aimant, & indiquer par-
là les courbes magnétiques qu'on
cherche à la mer, sur les cartes
réduites. xxxi, 168, xxii p.
2 maps. Paris, 1776.
Mp178, 232; R; W459.   APS

308. LE ROY, JEAN-BAPTISTE, d.
        1800. Description de l'ap-
pereil qui paroit le plus propre
pour faire des observations sur
l'électricité de l'air, des nuées
orageuses & de la foudre. 7p. 1pl.
n.p., n.d.
Benjamin Franklin's copy.
Not in MRW.   HSP

309. ----. Mémoire sur la forme
    des barres ou des conducteurs
métalliques, destinés à préserver
les édifices des effets de la
foudre, en transmettant son feu à
la terre... 16p. 1pl. [Paris,
1777.]
Benjamin Franklin's copy.
Not in MRW.   APS

310. LICHTENBERG, GEORG CHRISTOPH,
    1742-1799. De nova methodo
naturam ac motum fluidi eletrici
investigandi. [2 commentaries]
15, 16p. 6pl. Göttingen, 1778-79.
Mp627; R; W482.   LCP, APS

see also: 150

311. [LICHTENBERG, LUDWIG CHRIS-
    TIAN], 1738-1812. Verhal-
tungs-Regeln bey nahen Donner-
wetter nebst den Mitteln sich gegen die schädlichen Wirkungen des
Blizes in Sicherheit zu setzen...
viii, 46, [2]p. illus., 1pl.
Gotha, 1774.
R.   FI, APS

LINNAEUS, CAROLUS: see 581

312. LOBE, WILHELMUS.
    De vi corporum electrica...
29, [2]p. Leyden, 1743.
Inaugural dissertation.
Mp555; R.   UP

LONGINUS: see GOCLENIUS

313. LORIMER, JOHN, 1732-1795.
    A concise essay on magnetism;
with an account of the declination
and inclination of the magnetic
needle; and an attempt to ascertain
the cause of the variation thereof.
xv, 34p. 6pl. port. London, 1795.
Inscription: from the author to
the A.P.S.
Mp30, 243; R; W594.   FI, LCP, APS

314. LOUIS, ANTOINE, 1723-1792.
    Lettre à M. Nollet...March
10, 1749. 19p. [Paris], 1749.
Not in MRW.   CP, APS

315. ----. Observations sur
    l'électricité, où l'on tâche
d'expliquer son méchanisme & ses
effets sur l'oeconomie animale avec
des remarques sur son usage. xxiv,
175, [3]p. Paris, 1747.
Mp186; R; W341.   APS, CP

316. LOVETT, RICHARD, 1692-1780.
    Electrical philosopher, con-
taining a new system of physics...
disposed in the form of a dialogue.
290p. Worcester, 1774.
Mp213; R; W447.   FI

317. ----. Philosophical essays
    in three parts. Containing
I. An inquiry into the nature and
properties of the electrical fluid
...II. A dissertation on the nature
of fire...III. A miscellaneous discourse...To which is subjoin'd, by
way of appendix, a clear and con-
cise account of the variation of
the magnetic needle...xxiv, [5],
525, [44]p. 4pl. Worcester, 1766.
Mp213; R; W417.   LCP

318. ----. The subtil medium
    prov'd: or, That wonderful
power of nature, so long ago con-
jectur'd by the most ancient and
remarkable philosophers, which
they call'd sometimes Aether, but
oftener elementary fire, verify'd
... 141, [5]p. London, 1756.
Mp212; R; W391.   LCP

319. LOWNDES, FRANCIS.
    Observations on medical elec-
tricity, containing a synopsis of
all the diseases in which electricity has been recommended or applied with success; likewise,
pointing out a new and more effi-
cacious method of applying this
remedy, by electric vibrations.
51p. London, 1787.
Mp629; R; W543.   LCP

320. ----. The utility of medical
    electricity illustrated, in
a series of cases, and practical
observations: tending to prove the

superiority of vibrations to every other mode of applying the electric fluid. 46p. London, 1791.
R.   CP

321.   LOZERAN DU FESC, LOUIS ANTOINE DE, d. 1755. Dissertation sur la cause et la nature du tonnerre et des éclairs. Avec l'explication des divers phénomènes qui en dépendent...Et lettre de l'auteur à M. Sarrau... [6], 100, viii p. Paris, 1727.
(R).   APS

322.   LUDWIG, CHRISTIAN, 1749-1784. De magnetismo in corpore humano. 40p. Leipzig, 1772.
Inaugural dissertation.
Not in MRW.   CP

323.   ----. Dissertatio de aethere varie moto causa diversitatis luminum. [3], 40, [1]p. 1pl. Leipzig, [1773].
Not in MRW.   APS

324.   LÜDERUS, GERHARDUS. ...De methodis demonstrandi declinationem magnetis variam et inconstantem...prima dissertatione... 56p. 1pl. Wittenberg, [1718].
Mp554; R; W247.   APS

325.   LUGT, HENDRIK. De theorie der electriciteit, rustende op proefondervindlyke waarheden. xii, 120p. 2pl. West-Zaandam, 1797.
Not in MRW.   FI

326.   ----. Onderwys in de erste beginzels der electriciteit; geschikt tot eene handleiding voor de zoodanigen die zich in deze wetenschap willen oefenen. 120p. 2pl. West-Zaandam, 1797.
Not in MRW.   APS

327.   LULL, RAMÓN, 1235?-1315. De virtute magnetis. [16]p. [Rome, ca. 1515].
Photocopy.
Not in MRW.   UP

328.   LULLIN, AMADEUS. Dissertatio physica de electricitate, quam, favente deo, praeside D.D. Hor. Ben. de Saussure... 55p. Geneva, 1766.
Mp226; R.   LCP

329.   LUYTS, JAN, 1655-1721, praeses. De magnete... 16p. Rhenum, 1716.
Inaugural dissertation.
Not in MRW.   UP

330.   LUZ, JOHANN FRIEDRICH, 1744-1827. Unterricht vom Blitz und den Blitz-oder Wetterableitern ... [3], 150p. 1pl. Frankfurt & Leipzig, 1784.
(R).   APS

331.   LYON, JOHN, 1734-1817. An account of several new phenomena, discovered in examining the bodies of a man and four horses, killed by lightning, near Dover, in Kent. With remarks on the insufficiency of the popular theory of electricity to explain them. 38p. London, 1796.
R.   LCP

332.   ----. Experiments and observations made with a view to point out the errors of the present received theory of electricity... xxiv, 280, [8]p. 2pl. London, 1780.
Mp281; R; W493.   FI, APS

333.   MACKAY, ANDREW, 1760-1809. The theory and practice of finding the longitude at sea or land: to which are added, various methods of determining the latitude of a place, and variation of the compass; with new tables. 2v. in 1. London, 1793.
Advertisement inserted.
Also 2d & 3rd eds.: Aberdeen, 1801 and London, 1810.
(W706).   APS

334.   MAFFEI, FRANCESCO SCIPIONE, MARQUESE DE, 1675-1755. Della formazione de'fulmini... anche degl'insetti regenerantisi,

e de'pesci di mare su i monti, e
più lungo dell'elettricità. [8],
189, [6]p. Verona, 1747.
Mp321, 555; R.   LCP, APS

see also: 122

335. MAGALOTTI, LORENZO, 1637-1712.
       Lettere scientifiche, ed eru-
dite. xxiv, 303p. port. Florence,
1721.
R; W253.   APS

see also: 164

336. MAGGIOTTO, FRANCESCO.
       Saggi sopra l'attivita del-
la macchina elettrica costrutta da
Francesco Maggiotto...ed alcuni
riflessi intorno l'elettrico fluido.
28p. lpl. Venice, 1781.
R.   LCP, APS

MAHON: see STANHOPE

337. MAIRAN, JEAN JACQUES DORTOUS
       DE, 1678-1771. Traité phy-
sique et historique de l'aurore
boréale. Suite de Mémoires de
l'Académie royale des sciences...
2d ed... [12], 570, xxii p. 17pl.
Paris, 1754.
Benjamin Franklin's copy.
Bookplate: Samuel Vaughan, Jr.
R; W382.   APS

338. MAKO, PAUL, 1723-1793.
       Physikalische Abhandlung von
den Eigenschaften des Donners und
den Mitteln wider das Einschlagen.
Verfasst von Paul Mako...und von
Joseph Edlen von Retzer seinem
Zuhörer in das Deutsche übersetzt.
[6], 125p. front. Vienna, 1772.
R.   LCP, APS

339. ----. Physikalische Abhand-
       lung... [10], 125, 44p.
front. Vienna, 1772.
"Gegenstände der Prüfung..." 44p.
at end.
R.   APS

340. [MANGIN,   ABBÉ], d. 1772.
       Histoire générale et parti-

culiere de l'électricité, ou ce
qu'on dit de curieux & d'amusant,
d'utile & d'intéressant, de ré-
jouissant & de badin, quelques
physiciens de l'Europe. 3v. lpl.
Paris, 1752.
Mp633; R; W372.   LCP, APS

341. MANIERA PRATICA di fare li
       conduttori ai campanili, alle
chiese, ed alle case, descritta per
uso dei fabbri, falegnami, e mura-
tori, e per la gente delle ville,
e del popolo. 37p. [Venice], 1787.
R.   LCP

342. MARAT, JEAN PAUL, 1743-1793.
       Découvertes sur le feu,
l'électricité et la lumière, con-
statées sur une suite d'expériences
nouvelles qui viennent d'être veri-
fiées par MM. les commissaires de
l'Academie des sciences. [4], 38p.
Paris, 1779.
Mp269; W483.   UP

343. ----. Découvertes... 2d ed.
       [4], 38p. Paris, 1779.
Mp269; R; W483.   LCP

344. ----. Mémoire sur l'élec-
       tricité medicale... 8, 111p.
Paris, 1784.
Mp269, 385; R; W524.   LCP, APS

345. ----. Physische Untersuch-
       nungen über die Elektricität
...Aus dem französischen übersetzt
mit Anmerkungen vom Christ. Ehrenfr.
Weigel. [10], 660p. 5pl. Leipzig,
1784.
R.   CP

346. ----. Recherches physiques
       sur l'électricité. viii,
461p. 5pl. Paris, 1782.
Mp269; R; W509.   FI, APS

347. MARHERR, PHILIPP AMBROSIUS,
       1738-1771. Abhandlung von
der Wirkung der Luftelektricität
in den menschlichen Körper...aus
dem Lateinischen übersetzt. [237]-
270p. [Chur & Lindau, 1770].
With F. Bauer's Experimental-

Abhandlung...(q.v.).
R.   APS

348. [MARIOTTI, PROSPERO], 1703-
      1767. Lettera scritta ad
una dama...sopra la cagione de'
fenomeni della macchina elettrica.
26p. [Perugia, 1748].
R.   APS

349. MARTIN, BENJAMIN, 1704-1782.
      Essai sur l'électricité con-
tenant des recherches sur sa nature, ses causes et propriétés,
fondées sur la théorie du mouvement de vibration...de M. Newton
...tr. de l'anglois. [55]-112p.
2pl. [Paris, 1748].
In Recueil de traités sur l'électricité...
Not in MRW.   APS

350. ----. An essay on electri-
      city; being an enquiry in-
to the nature, cause and properties thereof, on the principles
of Sir Isaac Newton's theory of
vibrating motion, light and fire;
and the various phaenomena of
forty-two capital experiments;
with some observations relative
to the uses that may be made of
this wonderful power of nature.
40p. Bath, 1746.
Benjamin Franklin's copy. (HSP)
Mp252; R; W327.   HSP, CP, APS

351. ----. Philosophia Britan-
      nica: or A new and compre-
hensive system of the Newtonian
philosophy, astronomy and geog-
raphy. In a course of twelve lectures, with notes... 2v. plates.
Reading, 1747.
(Mp252); W342.   APS

352. ----. A supplement to the
      Philosophia britannica. Ap-
pendix I. Containing new experiments in electricity, and the method of making artificial magnets.
32p. plates. London, 1759.
Also Appendix II on optical instruments pp.35-80.
R; W342a.   FI

353. MARTIUS, JOHANN NICOLAUS.
      De magia naturali, ejusque
usu medico ad magice et magica
curandum... 44p. Erfurt, 1700.
Inaugural dissertation.
W224.   CP

354. MARUM, MARTIN VAN, 1750-1837.
      Beschreibung einer ungemein
grossen Elektrisier-Maschine und
der damit im Teylerschen Museum zu
Haarlem angestelten Versuche...aus
dem Holländischen übersetzt. 2v.
plates. Leipzig, 1786-88.
R.   APS

355. ----. Beschryving eener on-
      gemeen groote electrizeer-
machine, geplaatst in Teyler's museum te Haarlem, en van de proefneemingen met dezelve in 't werk
gesteld. xxxi, 205p. 6pl.
Haarlem, 1785.
Dutch and French text.
R; W532.   LCP

356. ----. Description de quel-
      ques appareils chimiques nou-
veaux ou perfectionnés de la Fonda-
tion Teylerienne, et des expériences faites avec ces appareils...
viii, 116, viii, 123p. 15pl. Haarlem, 1798.
French and Dutch text.
Inscription from the author to
Thomas P. Smith.
R.   APS

357. ----. Eerste vervolg der
      proefneemingen, gedaan met
Teyler's electrizeer-machine. xix,
266, 11p. 12pl. Haarlem, 1787.
"Aanhangzel..." pp. 233-66.
"Byvoegzel tot de beschryving..."
11p. at end.
Dutch and French text.
R; W532.   LCP

358. ----. Second continuation
      des expériences faites par
le moyen de la machine électrique
Teylerienne. xv, 391p. plates.
Haarlem, 1795.
Text in French and Dutch.
Dutch title: Tweede vervolg der

Proefneemingen gedaan met Teyler's electrizeer-machine.
R.   APS

359. ----. Sur la théorie de
    Franklin, suivant lequel les phénomènes électriques sont expliqués par un seul fluide. 29p. illus. lpl. [Haarlem, 1819].
W761.   APS

360. MASARS DE CAZÈLES,
    Mémoire sur l'électricité médicale, et histoire du traitement de vingt malades traités, et la plupart guéris par l'électricité. 122p. Paris & Toulouse, [1780].
Mp229; R.   LCP

361. MAUDUYT DE LA VARENNE, PIERRE
    JEAN ÉTIENNE. Extraits des journaux tenus pour quatre-vingt-deux malades qui ont étés élec-trisés... 49p. 1 table. Paris, 1779.
Not in MRW.   LCP

362. ----. Mémoire sur les dif-
    férentes manières d'adminis-trer l'électricité, et observations sur l'effets qu'elles ont produits... 30lp. 2pl. Paris, 1784.
Extrait des Memoires de la Société royale de médecine.
Mp263; R.   LCP, APS

363. ----. Memoria sobre los
    diferentes modos de adminis-trar la electricidad; y observaciones sobre los efectos que estos diversos modos han producido:... tr. en castellano por el Capitan D. Vincente Alcalá-Galiano.
xviii, 210p. 2pl. Segovia, 1786.
Not in MRW.   LCP

364. MAZZACANE, CARLO.
    Lettere sopra lo elettri-cismo...a S.E. il signor Marchese Andreasi... [4], 159p.
Naples, 1780.
R.   APS

365. MAZZOLARI, GIUSEPPE MARIA,

1712-1786. ...Electricorum, libri VI. 188 [i.e. 288]p. 2pl.
Rome, 1767.
R.   APS

MESMER, FRIEDRICH ANTON: see 507,
    1033, 1068, 1216

366. MICHELL, JOHN, 1724-1793.
    Traité sur les aimans arti-ficiels; contenant une méthode courte & aisée pour les composer ...tr. de deux ouvrages anglois de J. Michell & J. Canton, par le P. Rivoire de la C. de J. Avec une préface historique du traducteur, où l'on expose les méthodes & les expériences de MM. Duhamel & Antheaume...pour perfectionner les aimans. vii, cxx, 160p. 4pl.
Paris, 1752.
Mp253; R; W358b.   LCP, APS

367. ----. A treatise of arti-
    ficial magnets; in which is shewn an easy and expeditious method of making them, superior to the best natural ones. And also, a way of improving the natural ones, and of changing or converting their poles. Directions are likewise given for making the mariner's nee-dles... 8lp. plate. Cambridge, 1750.
Mp191; R; W358.   LCP, APS

368. ----. A treatise of arti-
    ficial magnets;...2d ed.
78p. lpl. Cambridge, 1751.
In vol. of pamphlets owned by Joseph Priestley. (LC)
Benjamin Franklin's copy. (HSP)
R; W358a.   LCP, HSP

369. MILLS, JOHN, d. 1784?
    An essay on the weather; with remarks on the shepherd of Banbury's rules for judging of it's changes; and directions for preserving lives and buildings from the fatal effects of lightening. Intended chiefly for the use of husbandmen. 2d ed. xv, [l], 127, viii p. London, 1773.
(R); W442.   APS

370. MILNER, THOMAS, 1719-1797.
Experiments & observations
in electricity. 2, vii-xvi, 111p.
2pl. London, 1783.
Mp367, 556; R; W517.   LCP, APS

MITCHELL, JOHN: see 509

MONTANARI, GIUSEPPE IGNAZIO: see
537

371. MORGAN, GEORGE CARDOGAN,
1754-1798. Lectures on
electricity. 2v. Norwich, 1794.
Mp282; R; W589.   FI, LCP

372. MORIN, JEAN, 1700-1764.
Nouvelle dissertation sur
l'électricité des corps, dans la-
quelle on develope le vrai mecan-
isme des plus suprenans phénomènes
... 200, [15]p. Chartres, 1748.
Mp187; R.   LCP, APS

373. MURHARD, FRIEDRICH WILHELM
AUGUST, 1779-1853. Versuch
einer historisch-chronologischen
Bibliographie des Magnetismus.
166p. Cassel, 1797.
Mp639; R.   APS

374. MUSCIO, GIAN GAETANO DEL.
Dissertazione...con cui si
risponde a vari dubbi promossi
contro la teoria dell'elettricismo del Franklin dal dottor
Giuseppe Saverio Poli nelle sue
Riflessioni intorno agli effeti
di alcuni fulmini. 66p. Naples,
1774.
R.   APS

375. MUSSCHENBROEK, PETRUS VAN,
1692-1761. Cours de phy-
sique experimentale et mathema-
tique,...tr. par M. Sigaud de la
Fond. 3v. plates. Paris, 1769.
Ms. notes (APS)
Mp174; W427.   LCP, APS

376. ----. Dissertatio physica
experimentalis de magnete.
Lugduni Batavorum anno MDCCXXIX
edita nunc vero auditoribus ob-
lata... 283p. 10pl. Vienna,
1754.
Mp175; R; W383.   APS

377. ----. Essai de physique...
avec une description de nouvelles sortes de machines pneuma-
tiques et un recueil d'expériences
par Mr. J.V.M. Tr. du hollandois
par Mr. Pierre Massuet. 2v. plates,
port. Leyden, 1751.
Mp639; R; W300.   APS

378. ----. Introductio ad philo-
sophiam naturalem. 2v. [20],
1132, [13]p. plates. Leyden, 1762.
Autograph: J. Vaughan, 1776.
R; (W312c).   LCP, APS

379. ----. ...Physicae experi-
mentales, et geometricae, de
magnete, tuborum capillarium vit-
reorumque speculorum attractione,
magnitudine terrae,...disserta-
tiones: ut et Ephemerides meteoro-
logicae ultrajectinae. [5], 685p.
28pl. Leyden, 1729.
R; W268.   APS

380. ----. Physicae experimen-
tales, et geometricae...
v.p. plates. Vienna, Prague, &
Tergesti, 1756.
p. 1-270 De Magnete...
R.   LCP

381. ----. Institutiones physicae conscriptae in usus
academicos. [8], 743, 14p. plates.
Leyden, 1748.
R; W312a.   LCP, CP

see also: 164

382. NAIRNE, EDWARD, 1726-1806.
The description and use of
Nairne's patent electrical machine; with the addition of some
philosophical experiments and medical observations. 68p. 5pl. London,
1783.
Joseph Priestley's copy.
Mp265; R; W518.   LCP

383. ----. Description de la machine électrique negative et

positive de M. Nairne, avec les
détails de ses applications à
la physique & principalement à la
médecine. Tr...par M. Caullet de
Veaumorel. xlvii, 179p. 5pl.
Paris, 1784.
Mp265.   LCP

384. [-----]. Directions for
      using the electrical ma-
chine as made & sold by Edward
Nairne... 11, [1]p. 1pl. [Lon-
don], 1764.
(W443).   FI, APS

385. ----. Electrical experi-
      ments...Read at the Royal
Society... 12p. 2pl. [London,
1774].
No t.p.
Benjamin Franklin's copy.
R.   HSP

386. ----. Experiments on elec-
      tricity, being an attempt to
shew the advantage of elevated
pointed conductors... 40p. 4pl.
London, 1779.
From Phil. Trans.
Benjamin Franklin's copy.
Mp252; R; W484.   APS

see also: 61, 63, 253

387. NEALE, JOHN.
      Directions for gentlemen who
have electrical machines... 77p.
plates. London, 1747.
R; W343.   FI

388. NECKERMANN, MICHAEL.
      Theoria ignis electrici una
cum selectis positionibus ex physica
experimentali...propugnabunt ornati
ac eruditi domini Michael Neckermann [et] Josephus Boggia. 96p.
Würzburg, 1770.
Not in MRW.   APS

389. [NEEDHAM, JOHN TURBERVILLE],
      1713-1781. A letter from
Paris, concerning some new electrical experiments made there.
7p. London, 1746.
Ms. notes by James Logan (HSP)

R; W354.   LCP, HSP, APS

see also: 497

390. NEGRO, SALVATORE DAL, 1768-
      1839. Dell'elettricismo
idro-metallico. 2 parts. viii,
116, xi, 84p. 1pl. Padua, 1802-
03.
R; W638.   APS

391. ----. Nuovo metodo di
      costruire macchine elettri-
che di grandezza illimitata e
nuovi sperimenti diretti a retti-
ficare l'apparato elettrico...
xxii, 112p. 1pl. Venice, 1799.
Mp589; R.   APS

392. ----. Terza serie di es-
      perimenti che servono di
fondamento ad un principio che
rende ragione delle nuove proprie-
tà degli elettromotori del Volta.
15p. Padua, 1834.
From Annali delle scienze del
Regno Lombardo-Veneto.
R.   UP

see also: 854

393. NICHOLSON, WILLIAM, 1753-
      1815. A dictionary of prac-
tical and theoretical chemistry,
with its application to the arts
and manufactures... 1v. unpaged.
plates. [London], 1808.
Appendix on chemical electricity
by Humphry Davy, 15p. (The Baker-
ian lecture)
Not in MRW.   UP

394. ----. An introduction to
      natural philosophy...3rd ed.
2v. 570, [8]p. 25pl. Philadelphia,
1788.
Also New ed., Phila., 1793-95.
(R); (510).   APS

395. NOLLET, JEAN ANTOINE, 1700-
      1770. Ensayo sobre la elec-
tricidad de los cuerpos...Tr. en
Castellano por D. Joseph Vazquez
y Morales. Añadida la Historia de
la electricidad. [24], lxvii,

[9], 131p. 1pl. Madrid, 1747.
Mp555; R.   UP, LCP, APS

396. ----. Essai sur l'électri-
     cité des corps. xx, [4],
227p. 4pl. front. Paris, 1746.
Ms. notes (APS).
Mp181; R; W329.   LCP, APS

397. ----. Essai sur l'électri-
     cité des corps. 2d ed.
xxiii, 272p. plates. Paris,
1750.
R.   FI

398. ----. Essai sur l'électri-
     cité des corps. 2d ed.
xxiii, 276p. plates, front.
Paris, 1765.
Not in MRW.   LCP

399. ----. Essai sur l'électri-
     cité des corps. 3rd ed.
xxiii, 273p. 4pl. front. Paris,
1754.
W329a.   APS

400. ----. Leçons de physique
     experimentale. 6v. plates,
front. Amsterdam & Paris, 1745-64.
(Mp182); R; (W319).   APS

401. ----. Leçons de physique
     expérimentale. 8th ed. 6v.
116pl. port. Paris, 1775.
v.5 is 6th ed., Paris, 1777.
Autograph: John Vaughan.
Not in MRW.   APS

402. ----. Leçons de physique
     expérimentale. 6v. plates,
port. Paris, 1777-83.
v.1-2 are 9th ed.; v.3 is 7th ed.;
v.4-5 are 6th ed.; v.6 is 8th ed.
(W319).   LCP

403. ----. Lettere intorno
     all'elettricità nelle quale
si esaminano le ultime scoperte
fatte in tal materia, e le conse-
guenze che dedur se ne possono.
180p. 4pl. Venice, 1755.
Not in MRW.   LCP, APS

404. ----. Lettres sur l'élec-
     tricité; dans lesquelles on

examine les dernières découvertes
qui ont été faites sur cette matiere, & les consequences que l'on
en peut tirer. xii, 264p. 4pl.
Paris, 1753.
R; W379.   FI, CP, APS

405. ----. Lettres sur l'élec-
     tricité. Nouvelle éd.
3 parts. plates. Paris, 1760-67.
R; W379a.   APS

406. ----. Lettres sur l'élec-
     tricité...Nouvelle éd. 3v.
plates. Paris, 1774-77. v.2 is 2d
ed., v.3 is 3rd ed.
Stamp: H. Cavendish (UP).
Not in MRW.   UP, APS

407. ----. Lettres sur l'élec-
     tricité. Nouvelle éd.
3 parts. plates. Paris, 1774-77.
Pt. 3 lacking.
Not in MRW.   APS

408. ----. Lettres sur l'élec-
     tricité, dans lesquelles on
examine les découvertes qui ont
été faites sur cette matiere depuis l'année 1752...Nouvelle éd.
2v. plates. Paris, 1764.
Not in MRW.   LCP

409. ----. Programme ou idée
     générale d'un cours de phy-
sique expérimentale, avec un catalogue raisonné des instrumens qui
servent aux expériences. xxxix,
[1], 190, [5]p. Paris, 1738.
R.   APS

410. ----. Recherches sur les
     causes particulières des
phénomènes électrique, et sur les
effets nuisibles ou avantageux
qu'on peut en attendre. xxxvi,
444p. 8pl. Paris, 1749.
Mp168, etc.; R; W355.   FI, APS,
                          LCP

411. ----. Recherches sur les
     causes particulières...
Nouvelle éd. xxxvi, 444p. plates.
Paris, 1754.
W355a.   LCP, FI

412. ----, Ricerche sopra le
     cause particolari de'fe-
nomeni elettrici e sopra gli effetti nocivi o vantaggiosi che se
ne può attendere. 334p. plates.
Venice, 1750.
R.   LCP

413. ----. Saggio intorno all'
     elettricità de'corpi...Tr.
dal francese. Aggiuntevi alcune
esperienze ed osservazioni che
illustrano l'istessa materia, del
sig. Guglielmo Watson. 254p. 5pl.
front. Venice, 1747.
Front missing (APS).
R.   LCP, APS

see also: 314

414. NORMAN, ROBERT, fl. 1590.
     The newe attractive, shewing
the nature, propertie, and manifold vertues of the loadstone;
with the declination of the needle,
touched therewith, under the plaine
of the horizon. iv, [2], 43p.
illus. London, 1720 (reprint).
First published 1581.
(Mp75); R; W66a.   APS

NOYA CARAFA, DUC DE, pseud.: see
     ADANSON

415. NUOVI RAGGUAGLI dell'es-
     perienze d'elettrometria or-
ganica eseguite in Brescia, Udine,
e Verona nell'anno MDCCXCIII.
viii, 88p. 1pl. Venice, 1794.
Not in MRW.   APS

416. OBSERVATIONS SUR l'électri-
     cité naturelle, par le moyen
d'un cerf-volant. Adressées à
l'Académie des sciences de Peters-
bourg. [Par] V.T. & T.O.S. 6p.
1pl. The Hague, 1778.
Not in MRW.   APS

417. O'GALLAGHER, FELIX.
     An essay on the investiga-
tion of the first principles of
nature: together with the application thereof to solve the phenomena of the physical system.

2v. 3pl. London, 1784-86.
Not in MRW.   APS

418. OLIVER, ANDREW. 1731-1799.
     An essay on comets, in two
parts... vi, 87p. 1pl. Salem,
Mass., 1772.
Not in MRW.   APS

see also: 1270-71

419. OZANAM, JACQUES, 1640-1717.
     Recreations mathematiques
et physiques, qui contiennent
plusieurs problèmes...Nouvelle éd.
4v. illus. plates. Paris, 1735-36.
Also London ed. 4v. 1814.
Mp401; R; W304.   APS

PARTHENIUS, JOSEPH MARIANUS: see
     MAZZOLARI

420. [PAULIAN, AIMÉ HENRI], 1722-
     1801. L'électricité soumise
a un nouvel examen, dans différentes lettres addressées à l'abbé
Nollet...par l'auteur du Diction-
naire de physique. xlviii, 286,
[2]p. 2pl. Avignon, 1768.
Mp555; R; W421.   APS

421. PEART, EDWARD, 1756-1824.
     On electric atmospheres. In
which the absurdity of positive &
negative electricity is incontest-
ably proved:...to which is prefixed a letter addressed to Mr.
Read of Knightsbridge; in reply to
his remarks on the Author's former
Tract on electricity, &c. 8, xliv,
81p. Gainsborough, 1793.
Ms. notes, signed J. Read.
R; W584.   APS

422. ----. On electricity; with
     occasional observations on
magnetism. Pointing out the in-
consistency & fallacy of the doctrine
of positive & negative electricity:
and investigating & explaining the
true principles, composition, &
properties of electric atmospheres.
91p. Gainsborough, 1791.
Mp556; R.   APS

3.

THE

Newe Attractive,

SHEWING

The NATURE, PROPERTIE, and
manifold VERTUES of the

LOADSTONE;

WITH THE

Declination of the NEEDLE,

Touched therewith, under the Plaine
of the HORIZON.

Found out and Difcovered

By ROBERT NORMAN.

LONDON:

Reprinted in the Year MDCCXX.

423. PENROSE, FRANCIS, 1718-1798.
Essay on magnetism: or, An
endeavor to explain the various
properties and effects of the loadstone: together with the causes of
the same. 40p. Oxford, 1753.
Mp645; R; W380.   LCP

424. ----. Treatise on electricity: wherein its various
phaenomena are accounted for, and
the cause of the attraction and
gravitation of solids, assigned.
To which is added, a short account,
how the electrical effluvia act
upon the animal frame, and in
what disorders the same may probably be applied with success...
40p. illus. Oxford, 1752.
Mp645; R; W373.   LCP

425. PERKINS, BENJAMIN DOUGLAS,
1744-1810, ed. Experiments
with the metallic tractors...pub-
lished by the surgeons Herholdt
and Rafn...Copenhagen; tr. into
German...thence into the English
language by Mr. Charles Kampf-
muller: also reports of about one
hundred and fifty cases in England... xxiv, 355, [4]p. London, 1799.
Not in MRW.   CP

426. ----. Influence of metallic tractors on the human
body, in removing various painful
inflammatory diseases...lately
discovered by Dr. Perkins, of North
America;...and a new field of enquiry opened in the modern science
of galvanism, or animal electricity. 2d ed. xv, 99, [1]p.
London, 1799.
R.   CP

427. PICKEL, JOHANN GEORG, 1751-
1838. Experimenta physico-
medica de electricitate et calore
animali. [4], 133, [2]p. Würzburg, 1778.
Inaugural dissertation.
Mp249; R.   CP

428. PIVATI, GIOVANNI FRANCESCO,

1689-1764. Della elettri-
cità medica. Lettera...al celebre
signor Francesco Maria Zanotti.
33p. [Lucca, 1747].
Mp647; R.   LCP

429. PLAZ, ANTONIUS GULIELMUS.
De magnetismo et electrici-
tate fascini experte praefatus.
[Cum vita candidati C. J. Pfoten-
hauer.] 15p. Leipzig, 1779.
Inaugural dissertation.
Not in MRW.   CP

430. POHL, JOSEPH, 1705-1778.
Tentamen physico-experimen-
tale, in principiis peripateticis
fundatum, super phaenomenis electricitatis... [14], 196, [12]p.
Prague, 1747.
R.   UP

431. POLI, GIUSEPPE SAVERIO, 1746-
1825. La formazione del tu-
ono, della folgore, e di varie al-
tre meteore, spiegata giusta le
idee del Signor Franklin. 24,
152p. Naples, 1772.
Lacks pp. 141-152 (APS).
Mp199; R.   LCP, APS

432. POLINIÈRE, PIERRE, 1671-1734.
Expériences de physique.
2d ed. [8], 553, [23]p. 16pl.
Paris, 1718.
Mp148, 163; R; W248.   APS

433. PONCELET, POLYCARPE, ca.
1720-ca. 1780. La nature
dans la formation du tonnerre, et
la reproduction des etres vivans,
pour servir d'introduction aux
vrais principes de l'agriculture.
2v. in 1. 2pl. front. Paris, 1766.
Mp226; R; W418.   LCP, APS

434. POWER, HENRY, 1623-1668.
Experimental philosophy, in
three books: containing new experiments microscopical, mercurial,
magnetical... [20], 193p. illus.
4pl. London, 1664.
Separate t.p. for each section
dated 1663.
Mp649; R; W155.   LCP

435. PRÉVOST, PIERRE, 1751-1839.
De l'origine des forces mag-
netiques. xxiii, [1], 231p. 2pl.
Geneva & Paris, 1788.
Mp650; R; W547.   FI, APS

436. PRIESTLEY, JOSEPH, 1733-1804.
An account of a new electro-
meter, contrived by Mr. William
Henly, and of several experiments
made by him. In a letter...to Dr.
Franklin. 8p. 1pl. London, 1773.
Benjamin Franklin's copy. (HSP)
Mp228; R; W445.   HSP, APS

437. ----. A familiar introduction to the study of electricity. 51p. 4pl. London, 1768.
R; W422.   APS, LCP, HSP

438. ----. A familiar introduc-
tion... 2d ed. 85p. 1pl.
London, 1769.
R; W422a.   FI

439. ----. A familiar introduc-
tion... 3rd ed. 85p. 5pl.
London, 1777.
Autograph: John Vaughan; presented
by him 1784.
R.   APS

440. ----. Geschichte und gegen-
wartiger Zustand der Elektricität, nebst eigenthümlichen Versuchen. nach der zweyten Ausgabe
aus dem Englischen übersetzt und
mit Anmerckungen begleitet von D.
Johann Georg Krünitz. xxxii,
517p. 8pl. Berlin & Stralsund,
1772.
R.   APS, LCP, UP

441. ----. Heads of lectures on
a course of experimental
philosophy, particularly including
chemistry, delivered at the New
college in Hackney. xxviii, 180p.
London, 1794.
W590.   APS

442. ----. Histoire de l'électricité, tr. de l'anglois...
avec des notes critiques. 3v.
plates. Paris, 1771.

R; W453b.   UP, LCP, APS

443. ----. The history and present state of electricity,
with original experiments. [2],
xxxi, 736, [8]p. 8pl. London,
1767.
Mp227; R.   FI, UP, LCP, HSP, APS

444. ----. The history and present state of electricity...
2d ed. 2, xxxii, 712, iii, [7]p.
8pl. London, 1769.
Mp227; R.   FI, UP, LCP, APS

445. ----. The history and present state of electricity...
3rd ed. 2v. 8pl. London, 1775.
Mp227; R.   FI, UP, APS

446. ----. The history and pres-
ent state of electricity...
4th ed. xxxii, 691, [7]p. 8pl.
London, 1775.
Presented by John Vaughan, 1784. (APS)
Mp227; R; W453.   UP, APS

447. ----. The history and pres-
ent state of electricity...
5th ed. xxxii, 641, iii p. 13pl.
London, 1794.
Autograph: E. Newton Harvey. (APS)
Mp227; R.   UP, LCP, APS

see also: 937

448. PRINGLE, SIR JOHN, 1707-1782.
A discourse on the torpedo,
delivered at the anniversary meet-
ing of the Royal Society... 32p.
London, 1775.
R; W454.   APS

449. [PUGET, LOUIS DE], 1629-1709.
Lettres écrites a un philo-
sophe, sur le choix d'une hypo-
thése propre à expliquer les effets de l'aiman. 138p. [Lyon,
1699].
No t.p.
R; W220.   APS

450. QUINTINE, L'ABBÉ DE LA.
Dissertation sur le magnétisme des corps... [2], 42, [3]p.

illus. Bordeaux, 1732.
Mp651; R.   UP

451. RABIQUEAU, CHARLES A.
    Le spectacle du feu élé-
mentaire, ou, Cours d'électricité expérimentale. Où l'on trouve
l'explication, la cause & le méchanisme du feu dans son origine
...où l'on dévoile l'abus des
pointes pour détruire le tonnerre.
... [6], 14, 31-34, 296p. 10pl.
Paris, 1753.
Benjamin Franklin's copy. (APS)
Mp555; R; W381.   LCP, APS

452. RACKSTROW, B   .
    Miscellaneous observations
together with a collection of experiments on electricity... iv,
72p. plate. London, 1748.
Mp555; R; W351.   FI, LCP

453. [RANCY,   DE]   .
    Essai de physique, en forme
de lettres; à l'usage des jeunes
personnes...augmenté d'une lettre
sur l'aimant, de réflexions sur
l'électricité, & d'un petit traité
sur le planétaire. xii, 584p.
Paris, 1768.
Not in MRW.   APS

454. READ, JOHN.
    A summary view of the spon-
taneous electricity of the earth
and atmosphere; wherein the causes
of lightning and thunder, as well
as the constant electrification
of the clouds and vapours, sus-
pended in the air, are explained.
With some new experiments and observations...to which is subjoined the atmospherico-electri-
cal journal, kept during two years
... vi, 160p. 1pl. London, 1793.
Mp312; R; W585.   APS

REAMUR, RENÉ ANTOINE FERCHAULT
    DE: see 143

455. RECUEIL DE diverses cri-
    tiques et contre-critiques,
faites d'une lettre sur les coëffures à l'électricité ou coëffures

anti-foudroyantes. [33]p. n.p.,
1753.
Not in MRW.   APS

456. RECUEIL DE traités sur
    l'électricité tr. de l'alle-
mand et de l'anglois. 3v. illus.
plates. [Paris, 1748].
Contains essays by Freke, J. H.
Winkler, B. Martin, W. Watson,
(q.v.).
R.   APS

457. REIMARUS, JOHANN ALBRECHT
    HEINRICH, 1729-1814. Neure
Bemerkungen vom Blitze; dessen
Bahn, Wirkung, sichern und bequem-
en Ableitung: aus zuverlässigen
Wahrnemungen von Wetterschlägen
dargelegt. xii, 386p. 9pl.
Hamburg, 1794.
R.   FI, APS

458. ----. Die Ursache des Ein-
    schlagens vom Blitze, nebst
dessen natürlichen Abwendung von
unsern Gebäuden, aus zuverlässigen Erfahrung von Wetterschlägen
vor Augen gelegt. 128p. Langen-
salza, 1769.
R.   APS

459. RETZ, NOËL, 1758-1810.
    Fragmens sur l'électricité
humaine. Premier mémoire, contenant les motifs & les moyens d'aug-
menter & de diminuer le fluide
électrique du corps humain...Sec-
onde mémoire, contenant des recherches sur la cause de la mort
des personnes foudroyées, & sur
les moyens de se préserver de la
foudre. xii, 108, 5p. Amsterdam
and Paris, 1785.
R.   CP

RETZER, JOSEPH FRIEDRICH, FREIHERR
    VON: see 338-9

460. RIBRIGHT, GEORGE.
    A curious collection of ex-
periments to be performed on the
electrical machines, made by Geo.
Ribright & son. 24p. 2pl. London, 1779.
(R).   FI

461. RIDLEY, MARK, 1560-1624.
Magneticall animadversions.
Made by Marke Ridley, doctor in
physicke. Upon certaine magneticall advertisements, lately published, from Maister William Barlow. [2], 43p. London, 1617.
Mp97; R.   UP

RITTENHOUSE, DAVID: see 1272,
     1275-77

462. RITTER, JOHANN WILHELM, 1776-
1810. Beweis, dass ein be-
ständiger Galvanismus den Lebens-
process in dem Thierreich begleite.
Nebst neuen Versuchen und Bemerkungen über den Galvanismus. xx,
[4], 174p. 2pl. Weimar, 1798.
(Mp327); R.   APS

RIVOIRE, ANTOINE: see 366

463. RÖNNBERG, BERNHARD HEINRICH,
1712-1760. Vernünftige Gedanken von den Ursachen der Elektricitaet, wolte bey Gelegenheit
der elektrischen Versuche, welche
zur Bezeugung aller schuldigen
Verehrung, Hochachtung und Freund-
schaft in Gegenwart seiner hohen
Gönner, Befoerderer und geehrtsten
Freunden anzustellen Willens ist
zur gütigen Nachsicht mittheilen...
29p. Wissmar, [1746?].
R.   APS

464. ROHAULT, JACQUES, 1620-
1675. Tractatus physicus.
Gallice emissus et recens Lati-
nate donatus, per Th. Bonetum
D. M. Cum animadversionibus An-
tonii le Grand. 2 parts. 253,
289p. plates. London, 1682.
W171a.   APS

465. ROMAS, JACQUES DE, 1713-
1776. Mémoire, sur les
moyens de se garantir de la foudre dans les maisons; suivi d'une
lettre sur l'invention du cerf-
volant électrique, avec les pièces
justificatives de cette même
lettre. xxiv, 156, [4]p. 1pl.
front. Bordeaux, 1776.

Mp204; R.   FI, LCP, APS

466. RONAYNE, THOMAS.
A letter...to Benjamin Franklin...inclosing an account of some
observations on atmospherical electricity in regard of fogs, mists,
&c. With some remarks, communicated
by Mr. William Henly. 10p. plate.
London, 1772.
Benjamin Franklin's copy. (HSP)
Mp201, 238; R.   LCP, HSP, APS

467. SAMBUCETI, LUIGI MARIA.
La forza elettrica dell'
amore; Componimento poetico...
15, [1]p. Bologna, 1758.
Not in MRW.   APS

468. SANDEN, HEINRICH VON, 1672-
1728. Dissertatio de suc-
cino, electricorum principe...
pp.65-128. [Leipzig, 1746].
In Hausen, Novi profectus... (q.v.)
R; W309a.   APS

469. SANS,   ABBÉ.
Guérison de la paralysie,
par l'électricité... xxvii, [1],
234p. 4pl. front. Paris, 1778.
Presented by the author.
Mp229, 385; R.   APS

470. ----. Neue und durch die
Erfahrung vollkommen be-
stättigte Beweisen wie die von
einem Schlagfluss gelähmte Kranke
vermittelst der Electricität si-
cher und volkommen geheilet werden
können. Aus dem Französischen
übersetzt [von J. C. Thenn]. 251,
[2]p. 4pl. Augsburg, 1780.
R.   CP

471. SARRABAT, NICOLAS, 1698-1737.
Nouvelle hipothèse sur les
variations de l'eguille aimantée...
[2], 48p. 1pl. Bordeaux, 1727.
Mp167; R.   APS

472. SAUSSURE, HORACE BÉNÉDICT
DE, 1740-1799. Della maniera
di preservare gli edifici dal fulmine: informazione al popolo. 38p.
Venice, 1772.

Not in MRW.   FI

473. ----. Voyages dans les
        Alpes, précédés d'un essai
sur l'histoire naturelle des en-
virons de Geneve. 4v. plates.
Neuchatel, 1779-86.
v. 1 & 2 only.
Autograph: Benjamin Smith Barton.
Mp271; R.   APS

see also: 527

474. SCARELLA, GIOVANNI BATTISTA,
        1711-1779. De magnete libri
quatuor in duos tomos... 2v. in 1.
2pl. Brescia, 1759.
Mp657; R; W399.   APS

475. SCHÄFFER, JAKOB CHRISTIAN,
        1718-1790. Abbildung und
Beschreibung des beständigen Electricitätträgers. Wobey einige neue
Versuche und deren sonderbare Er-
folge Naturkündigern und Freunden
der Electricität zu genauerer Prüfung empfohlen werden. [1], 48p.
2pl. Regensburg, 1776.
R.   APS

476. ----. Abbildung und Beschrei-
        bung der electrischen Pistole
und eines kleinen zu Versuchen sehr
bequemen Electricitätträgers. Bey
welcher Gelegenheit zugleich von
einem Luftelectrophore vorläufige
Nachricht ertheilet wird. 32p. 3pl.
Regensburg, 1778.
Mp249; R.   APS

477. ----. Fernere Versuche mit
        dem beständigen Electrici-
tätträger. Nebst Beantwortung einiger dagegen gemachten Einwürfe.
56p. plate. Regensburg, 1777.

478. ----. Kräfte, Wirkungen
        und Bewegungsgesetze des be-
ständigen Electricitätträgers...
50p. 1pl. Regensburg, 1776.
Mp237; R.   APS

479. SCHAEFFER, JOHANN GOTTLIEB,
        1720-1795. Die electrische
Medicin, oder die Kraft und Wir-

kung der Electricität in dem
menschlichen Koerper und dessen
Krankheiten besonders bey gelähm-
ten Gliedern. [6], 84p. front.
Regensburg, 1766.
(R).   APS

480. SECONDAT, JEAN BAPTISTE,
        BARON DE, 1716-1796.
Mémoire sur l'électricité. 24p.
Paris, 1746.
R.   LCP

481. [----]. Mémoire sur l'élec-
        tricité. iv, 37p. Paris,
1746.
R.   FI, APS

482. [----]. Observations de
        physique et d'histoire
naturelle. 190p. n.p., n.d.
No t.p.
Joseph Priestley's copy.
Not in MRW.   UP

483. ----. Observations de phy-
        sique et d'histoire naturelle
sur les eaux minerales de Dax, de
Bagneres, & de Baregem...Histoire
de l'électricité, &c. [4], 205,
[3]p. Paris, 1750.
Mp131; R.   APS

484. [----]. Suite du memoire
        sur l'électricité. 30p.
Paris, 1748.
Not in MRW.   FI, APS

485. SEGNITZ, FRIEDRICH LUDWIG.
        De electricitate animali quam
dicere solent magnetismum animalem.
34p. Jena, 1790.
Inaugural dissertation.
Mp556; W566.   CP

486. [SGUARIO, EUSEBIO]. Dell'
        elettricismo: o sia della
forze elettriche de'corpi; svelate
dalla fisica sperimentale, con
un'ampia dichiarazione della luce
elettrica...aggiuntevi due disser-
tazioni attinenti all'uso medico
di tali forze. xvi, 391p. front.
Venice, 1746.
Mp385, 555; R.   LCP, APS

487. SIGAUD-LAFOND, JOSEPH AIGNAN, 1730-1810. Description et usage d'un cabinet de physique expérimentale...3. éd. revue, corrigée et augmentée par M. Rouland... 2v. plates. Tours, 1796.
(Mp280); (W455).   APS

488. ----. Elementos de fisica teórica y experimental... Traducidos. Añadiendo la descripcion de las máquinas y modo de hacer los experimentos;...Por D. Tadeo Lope. 6v. plates. Madrid, 1787-89.
(W543bis).   APS

489. ----. Examen de quelques principes erronés en électricité. 49p. Paris, 1796.
R.   APS

490. ----. Leçons de physique experimentale. 2v. 18pl. Paris, 1767.
W434.   APS, LCP

491. ----. Précis historique et expérimental des phénomènes électriques, depuis l'origine de cette découverte jusqu'à ce jour. xvi, 742, [2]p. 9pl. Paris, 1781.
R; W505.   UP, LCP, APS

492. ----. Précis historique et expérimental...2d ed. xvi, [4], 624, [4]p. 10pl. Paris, 1785.
Mp280; R; W505a.   FI, APS

493. ----. Traité de l'électricité. Dans lequel on expose, & on démontre par expérience, toutes les découvertes électriques, faites jusqu'à ce jour, pour servir de suite aux leçons de physique du même auteur. xxx, 413p. 12pl. Paris, 1771.
Mp385; R; W434.   LCP, APS

494. SOCIN, ABEL, 1729-1808. Anfangsgründe der Elektricität, in welchen hauptsächlich von den geriebenen elektrischen Körpern, der Elektricität welchen sie den unelektrischen mittheilen,

...gehandelt wird. In acht Vorlesungen... 124, [4]p. 1pl. Hanau, 1777.
R.   LCP, APS

495. SPALLANZANI, LAZZARO, 1729-1799. Lettera...al Sig. Marchese Lucchesini. 32p. [Pavia, 1783].
R.   APS

496. ----. Lettera...al Sig. Thouvenel sull'elettricità organica e minerale. 31p. Pavia, 1793.
R.   LCP

SQUARIO: see SGUARIO

497. STANHOPE, CHARLES, 3RD EARL, 1753-1816. Principes d'électricité, contenant plusieurs théorêmes appuyés par des expériences nouvelles, avec une analyse des avantages supérieurs des conducteurs élevés et pointus...Tr. de l'anglois par Mr. l'Abbé N... [John Turberville Needham]. 250p. plates. London & Brussels, 1781.
Mp255; R; W485a.   LCP

498. ----. Principles of electricity, containing divers new theorems and experiments, together with an analysis of the superior advantages of high and pointed conductors...by Charles viscount Mahon. xiv, 263p. 6pl. London, 1779.
Mp184; R; W485.   FI, LCP, APS

499. ----. Remarks on Mr. Brydone's account of a remarkable thunder-storm in Scotland... 23p. 1pl. London, 1787.
From Phil. Trans. v.77.
Not in MRW.   APS

500. STEAVENSON, ROBERT, 1756-1828. ...De electricitate et operatione ejus in morbis curandis. 35p. Edinburgh, 1778.
Inaugural dissertation.
Inscribed by author to Dr. Logan.
Mp556; W475.   LCP, CP

501. STEIGLEHNER, COELESTINO,
     1738-1819. Observationes
phaenomenorum electricorum in
Hohen-Gebrachin et prifling prope
ratisbonam factae et expositae...
55p. 1 table. [Regensburg], 1773.
Mp274; R.   APS

see also: 507

502. STRADA, FAMIANUS, 1572-1649.
     Prolusiones academicae...
[33], 564p. [Lyons, 1617].
No t.p.
Contains poem on imaginary mag-
netic telegraph.
(R); W90.   LCP

STUBBE: see STUBBS

503. STUBBS, HENRY, 1632-1676.
     Miraculous conformist: or,
An account of severall marvailous
cures performed by the stroaking
of the hands of Mr. Valentine
Greatarick;...in a letter to the
honourable Robert Boyle, Esq; with
a letter relating some other of
his miraculous cures... [4], 44p.
Oxford, 1666.
Not in MRW.   CP

504. STURM, JOHANN CHRISTOPH,
     1635-1703. Magnorum mundi
corporum magnetismus quem...prae-
sidio m. Joh. Christophorii Sturmii
... 20p. Altdorf, [1671].
Not in MRW.   UP

505. SULZER, JOHANN GEORG. 1720-
     1799. Nouvelle théorie des
plaisirs,...avec des réflexions
sur l'origine du plaisir, par Mr.
Kaestner. [2], 363p. 1pl. n.p.,
1767.
Mp223; R; W420.   APS

506. SWAMMERDAM, JAN, 1637-1680.
     Biblia naturae... 2v. 52pl.
Leyden, 1737-38.
W291.   CP, LCP

506a. SWEDENBORG, EMANUEL, 1688-
     1772. Principia rerum na-
turalium sive novorum tentaminum

phaenomena mundi elementaris
philosophice explicandi. [12],
452p. 27pl. port. Dresden &
Leipzig, 1734.
W283.   APS

507. SWINDEN, JAN HENDRIK VAN,
     1746-1823. Analogie de
l'électricité, et du magnétisme;
ou, Recueil de mémoires, couron-
nés par l'Académie de Baviere;
avec des notes et des disserta-
tions nouvelles. 3v. 8pl. The
Hague & Paris, 1785.
Mp272; R; W496a.   APS

508. SYMES, RICHARD.
     Fire analysed; or, the sev-
eral parts of which it is com-
pounded clearly demonstrated by
experiments. The teutonic philosophy proved true by the same
experiments. And the manner and
method of making electricity medi-
cinal and healing confirmed by a
variety of cures. vii, 87p.
Bristol, 1771.
Mp385; R; W435.   LCP

509. SYMMER, ROBERT, d. 1763.
     New experiments and obser-
vations concerning electricity.
In four papers read at...the
Royal society...With a letter from
John Mitchell...to Thomas Birch...
relating to some of those experi-
ments. 59p. London, 1760.
(R); W402.   APS

510. TANA, AGOSTINO.
     Elogio del padre Beccaria...
30p. Turin, 1781.
Not in MRW.   APS

511. THEATRUM SYMPATHETICUM auc-
     tum, exhibens varios authores.
De pulvere sympathetico...De un-
guento verò armorio... [4], 722,
[42]p. Nuremberg, 1662.
W152.   CP

THOURET, MICHEL AUGUSTIN: see 11,
     1043, 1072, 1116

512. [THOUVENEL, PIERRE], 1747-

44

1815. La guerra di dieci anni; raccolta polemico-fisica sull'elettrometria galvano-organica. Parte italiana-parte francese. 344p. 1pl. Verona, 1802.
R. APS

513. [----]. Mémoire physique et médicinal, montrant des reaports évidens entre les phénomenes de la baguette divinatoire, du magnétisme et de l'électricité... 304p. London & Paris, 1781.
Mp401; R; W506. APS

514. [----]. Seconde mémoire physique et médicinal montrant des rapports évidens entre les phénomenes de la baguette divinatoire, du magnétisme et de l'électricité. Avec des éclaircissemens sur d'autres objets non moins importans, qui y sont relatifs. Par M. T---, D.M.M. 268p. London & Paris, 1784.
W506a. APS

515. TITIUS, JOHANN DANIEL, 1727-1796. De electrici experimenti lugdunensis inventore primo ... 12p. Wittenberg, 1771.
Benjamin Franklin's copy.
Mp665; R. HSP

516. TOALDO, GIUSEPPE, 1719-1797 or '98. Breve difesa dei conduttori. (Giunta al Giornale astrometeorologico). 4p. [Padua, 1796].
Not in MRW. APS

517. ----. Dei conduttori per preservare gli edifizi da' fulmini. Memorie...Nuova ed. ritoccate ed accresciute di un appendice... xii, 104p. illus. Venice, 1778.
Mp253; R. LCP

518. [----]. Del conduttore elettrico posto nel campanile di S. Marco in Venezia. Memoria, in cui occasionalmente si ragiona dei conduttori, che possono applicarsi al vascelli,

al magazzini da polvere, ed altri edifizi. 37p. front. Venice, 1776.
R. LCP, APS

519. ----. Dell'uso de'conduttori metallici a preservazione degli edifizi contro de'fulmini, nuova apologia colla descrizione del conduttore della pubblica specola di Padova...con una lettera del sig. Franklin [al sig. di Saussure 8 oct. 1772]. 32p. 1pl. Venice, 1774.
R; W449. APS

520. ----. Memoires sur les conducteurs pour préserver les édifices de la foudre...tr. de l'italien avec des notes & des additions par Mr. Barbier de Tinan. x, 241p. 3pl. Strasbourg, 1779.
R; W449a. LCP, APS

see also: 85, 527

521. TODERINI, GIAMBATTISTA. Filosofia Frankliniana delle punte preservatrici dal fulmine, particolarmente applicata alle polveriere, alle navi, e a Santa Barbara in mare. 65p. Modena, 1771.
Benjamin Franklin's copy. (HSP)
R. LCP, HSP, APS

522. TRESSAN, LOUIS ELISABETH DE LA VERGNE, COMTE DE, 1705-1783. Essai sur le fluide électrique, considéré comme agent universel. 2v. Paris, 1786.
Mp190; R; W537. LCP, APS

523. TURINI, PIETRO. Considerazioni intorno all' elettricità delle nubi, ed al modo di applicare i conduttori alle fabbriche, e di preservare dal fulmine i depositi della polvere. 68p. Venice, 1780.
Benjamin Franklin's copy.
R; W495. HSP

524. UNZER, JOHANN CHRISTOPH,

1747-1809. Beschreibung eines mit den künstlichen Magneten angestellten medicinischen Versuchs. 144p. Hamburg, 1775.
Not in MRW.   APS

525. VALLEMONT, PIERRE LE LOR-
RAIN DE, 1649-1721. Description de l'aimant, qui s'est formé à la pointe du clocher neuf de N. Dame de Chartres: avec plusieurs expériences trés-curieuses, sur l'aimant & sur d'autres matiéres de physique. [6], 215p. illus. Paris, 1692.
Mp144; R; W205.   APS

526. VALLI, EUSEBIO, 1755-1816. Experiments on animal electricity, with their application to physiology. and some pathological and medical observations. xvi, 323p. London, 1793.
Autograph: James Rush
Mp302; R; W586.   LCP

VASQUEZ Y MORALES, JOSEPH:
      see 395

527. VASSALLI-EANDI, ANTONIO MARIA, 1761-1825. Lettere fisico-meteorologiche dei celeberissimi fisici Senebier, de Saussure, e Toaldo, con le risposte di A.-M. Vassalli. 223p. Turin, 1789.
R.   FI

528. ----. Recherches sur la nature du fluide galvanique. 31p. [Turin, 1805].
No t.p.
Inscription by author to Philip Tidyman; also latter's bookplate.
R.   APS

529. VAUGHAN, JOHN, 1775-1809. The valedictory lecture delivered before the Philosophical society of Delaware. 36p. Wilmington, 1800.
Presented by the author.
Not in MRW.   APS

530. VERATTI, GIUSEPPE, 1707-1793. Osservazioni fisico-mediche

intorno alla elettricità. [17], 143p. Bologna, 1748.
R.   APS

531. VIACINNA, CARLO. Del fulmine, e della sicura maniera di evitarne gli effetti; dialoghi tre di Carlo Viacinna a Matteo Allaghia. [4], 156p. Milan, 1766.
R.   LCP, APS

532. VILLENEUVE, OLIVIER DE. Essai de dissertation medico-physique sur les experiences de l'electricité... 22p. Paris, 1748.
Mp385; R.   APS

533. VOLLMAR, JOHANNES GUILLIEL-
MUS. De fulmine tactis. 24p. Strassburg, 1765.
Inaugural dissertation.
Not in MRW.   CP

534. VOLTA, ALESSANDRO GIUSEPPE
ANTONIO ANASTASIO, CONTE, 1745-1827. Collezione dell'opere ...[ed. by Vincenzio Antinori]. 3 parts in 5 vol. 7pl. port. Florence, 1816.
R; W731.   FI, UP, CP, APS

535. ----. De vi attractiva ignis electrici, ac phaenomenis inde pendentibus...ad Joannem Baptistam Beccariam. Dissertatio epistolaris. 72p. Como, 1769.
Joseph Priestley's copy.
Mp246; R; W428.   UP

536. ----. Lettere...sull'aria infiammabile nativa delle paludi. 147p. Milan, 1777.
R.   APS

537. ----. Lettere inedite di Alessandro Volta, coll'elogio del medesimo scritto del Prof. Pietro Configliacchi. [ed. by Giuseppe Ignazio Montanari]. 212p. 1pl. Pesaro, 1835.
R.   APS

538. ----. Lettres...sur l'air

46

inflammable des marais.
Auxquelles on a ajouté trois let-
tres du même auteur tirées du
Journal de Milan. Tr. de l'italien.
[6], 191p. 1pl. Strasbourg, 1778.
R.   APS

see also: 631

539. WAGNER, GODOFREDUS.
     De fulmine. [32]p. Witten-
berg, 1710.
Inaugural dissertation.
Not in MRW.   CP

540. WAKELEY, ANDREW.
     Mariners' compass recti-
fied... 275p. illus. London,
1780.
(W544).   FI

WALCKIERS: see 89

541. WALKER, RALPH.
     The memorial of Ralph Walker
to the honorable the Board of longitude. 18p. [London, 1794].
Autograph of author.
Not in MRW.   APS

542. ----. Treatise on magne-
          tism, with a description and
explanation of a meridional and
azimuth compass, for ascertaining
the quantity of variation...Also
improvements upon compasses in
general. With tables of variation, for all latitudes and lon-
gitudes. 226p. 7pl. London,
1794.
Inscription by author to Lewis
Dunbar. (c.1)
Autograph of author. (c.2)
R; W592.   APS

543. ----. A treatise on the
          magnet, or natural load-
stone, with tables of the variation of the magnetic needle...to-
gether with tables of the dip of
the needle...and a description of
a new-invented meridional and azimuth compass...To which is added
an appendix, containing hints to
ship-builders and navigators.

9, [2], [9]-226p. 7pl. London,
1798.
Mp669; R.   LCP

544. WALSH, JOHN, 1725?-1795.
     Of the electric property of
the torpedo. In a letter...to Ben-
jamin Franklin... 24p. plate.
London, 1774.
From Phil. Trans.
Benjamin Franklin's copy (HSP).
Presented by Franklin (APS).
R.   APS, HSP

545. ----. Of torpedos found on
          the coast of England. In a
letter to Thomas Pennant... 12p.
London, 1774.
Benjamin Franklin's copy.
R.   HSP

546. WATKINS, FRANCIS.
     A particular account of the
electrical experiments hitherto
made publick, with variety of new
ones...to which is annex'd, the description of a compleat electrical
machine... 77p. 3pl. London, 1747.
R.   LCP

547. WATSON, SIR WILLIAM, 1715-
          1787. An account of the ex-
periments made by some gentlemen
of the Royal society, in order to
discover whether the electrical
power would be sensible at great
distances. With an experimental
inquiry concerning the respective
velocities of electricity and
sound... 90p. London, 1748.
Mp176; W352.   LCP

548. ----. Expériences et ob-
          servations, pour servir à
l'explication de la nature et des
propriétés de l'électricité pro-
posés en trois lettres à la Société royale de Londres. tr. de
l'anglois d'après la 2ieme ed.
xii, 141, [3]p. Paris, 1748.
In Recueil de traités sur l'électricité...v.2.
Not in MRW.   UP, LCP, APS

549. ----. Experiments & obser-

vations tending to illus-
trate the nature & properties of
electricity. In one letter to
Martin Folkes, Esq; President, and
two to the Royal Society. 3rd ed.
viii, 59p. London, 1746.
Mp175; R; W333.   FI, LCP, APS

550. ----. A sequel to the Ex-
      periments and observations
tending to illustrate the nature
and properties of electricity...
80p. front. London, 1746.
Benjamin Franklin's copy (HSP).
Mp175; R; W333b.   FI, HSP, APS

551. ----. A sequel... 2d ed.
      80p. London, 1746.
(W333b).   LCP

see also: 413

552. WEBER, JOSEPH, 1753-1831.
      Beschreibung des Luftelek-
trophors. Nebst angehängten neuen
Erfahrungen, neuen Instrumenten,
einem Unterrichte von Zubereitung
der brennbaren Luft...Neueste,
mit der Beschreibung der elektrischen Lampe, vermehrte Auflage.
86p. 3pl. Augsburg, 1779.
R; W486.   LCP, APS

553. ----. Neue Erfahrungen
      idiolektrische Körper ohne
einigen Reiben zu elektrisiren.
[24], 118p. 3pl. Augsburg, 1781.
R.   LCP, APS

554. ----. ...Positiver Luft-
      elektrophor samt der And-
wendung desselben auf eine Elek-
trisirmachine. [10], 118p. 2pl.
Augsburg, 1782.
R.   APS

555. ----. Vollständige Lehre
      von den Gesetzen der Elek-
tricität und von der Anwendung
derselben... [28], 368p. 2pl.
Landshut, 1791.
R.   APS

556. ----. Vom dynamischen
      Leben der Natur überhaupt,

und vom **elektrischen** Leben im
Doppelelektrophor insbesondere.
151p. Landshut, 1816.
R; W732.   APS

see also: 1027

WEIGEL, CHRISTIAN EHRENFRIED:
      see 345

557. WHISTON, WILLIAM, 1667-1752.
      The longitude and latitude
found by the inclinatory or dipping
needle; wherein the laws of magnetism are also discover'd. To which
is prefix'd, an historical preface;
and to which is subjoin'd Mr. Robert
Norman's New attractive, or account
of the first invention of the dipping
needle. [2], xxviii, 115p. 2pl.
London, 1721.
New attractive...bound separately,
(q.v.).
Mp77; R; W256a.   APS

WIDENMANN, JOHANN FRIEDRICH WILHELM:
      see 220

558. WIEDEBURG, JOHANN ERNST
      BASILIUS, 1733-1789. Beo-
bachtungen und Muthmasungen über
die Nord-lichter. 96p. Jena,
1771.
Mp140; R.   FI

559. WILCKE, JOHANN CARL, 1732-
      1796. Amplissimi philosoph-
orum ordinis in Academia Rostoch-
iensi...Disputationem solemnem
philosophicam, De electricitatibus
contrariis... 12p. Rostock,
[1757].
Benjamin Franklin's copy.
Not in MRW.   HSP

560. ----. Disputatio physico
      experimentalis, de electri-
citatibus contrariis... 142p.
Rostock, [1757].
Autograph: Benjamin Franklin.
Ms. diagram tipped in after p.96.
Mp216; R.   HSP

561. ----. Tal, om de nyaste
      förklaringar öfver norr-

skenet... 110p. Stockholm, 1778.
Benjamin Franklin's copy.
R.   HSP

see also: 170

562. WILKINS, JOHN, BP., 1614-
     1672. Mercury: or, The
secret and swift messenger. Shewing, how a man may with privacy
and speed communicate his thoughts
to a friend at any distance. 3rd
ed. [8], 90p. London, 1707.
In the author's Mathematical and
philosophical works...1708.
Mp120; R;(W117).   APS

563. WILKINSON, CHARLES HENRY,
     fl. 1800. An analysis of a
course of lectures on the principles of natural philosophy...to
which is prefixed an essay on electricity. v, 220p. London, 1799.
Not in MRW.   LCP

564. ----. The effects of elec-
     tricity in paralytic and
rheumatic affections, gutta
serena, deafness, indurations of
the liver, dropsy, chlorosis, and
many other female complaints, &c.
...to which are added, some observations on the inefficacy of metallic tractors; and an analysis
of a course of lectures on experimental philosophy. 220p.
London, 1799.
R; W619.   LCP

565. ----. Elements of Galvan-
     ism, in theory and practice;
with a comprehensive view of its
history...Containing also, practical directions for constructing
the Galvanic apparatus, and plain
systematic instructions for per-
forming all the various experiments... 2v. plates, front.
London, 1804.
Mp140, etc.; R; W667.   FI, LCP
                         APS, CP

566. [WILSON, BENJAMIN], 1721-
     1788. An account of ex-
periments made at the Pantheon,

on the nature and use of conductors: to which are added, some
new experiments with the Leyden
phial... [2], 100p. 4pl. front.
London, 1778.
R; W478.   UP, LCP, APS

567. ----. An essay towards an
     explication of the phenom-
ena of electricity, deduced from
the aether of Sir Isaac Newton,
contained in three papers which
were read before the Royal society. xv, 95p. London, 1746.
Benjamin Franklin's copy (HSP).
Mp183; R; W334.   FI, LCP, APS,
                   HSP

568. ----. Further observations
     upon lightning; together
with some experiments. Communicated to the Royal society, and
rejected in the committee. vii,
26p. London, 1774.
R; W446a.   APS

569. ----. A letter...to Mr.
     Aepinus...read at the Royal
society... 35p. 1pl. London,
1764.
Not in MRW.   FI, LCP

570. ----. New experiments &
     observations on the nature &
use of conductors. 23p. [London,
1777].
Ms. notes.
Benjamin Franklin's copy.
Mp252; R.   APS

571. ----. A short view of elec-
     tricity. 37p. illus. 1pl.
London, 1780.
R.   LCP

572. ----. A treatise on elec-
     tricity. 2d ed. vii, iv,
224p. 5pl. port. London, 1752.
Benjamin Franklin's copy (CP).
Mp155, 185; R; W362a.   APS, CP

see also: 257-58

573. WINDLER A STORTEWAGEN, PETER
     JOHANN. Tentamina de causa

electricitatis quibus brevis historia nonnullus auctoribus, qui
hanc praecipue excoluerunt materi-
am, praemissa est. [8], 28p. 3pl.
Naples, 1747.
Not in MRW.   FI, APS

574.   WINKLER, JOHANN HEINRICH,
        1703-1770. De avertendi
fulminis artificio ex doctrina
electricitatis disserit. 20p.
1pl. Leipzig, [1753].
R.   APS

575.   ----. De vi luninsis bore-
        alis in commovenda acu mag-
netica. 8p. Leipzig, 1768.
Inaugural dissertation.
R.   CP

576.   ----. Die Eigenschaften der
        electrischen Materie und des
electrischen feuers aus verschie-
denen neuen Versuchen erklaeret,
und, nebst eklichen neuen Machinen
zum Electrisiren... [12], 164p.
4pl. Leipzig, 1745.
R; W323.   APS

577.   ----. Essai sur la nature,
        les effets et les causes de
l'electricite, avec une description de deux nouvelles machines...
tr. de l'allemand. vii, [3],
156p. 5pl. Paris, 1748.
In Recueil de traités de l'électricité...v.1.
R; W313c.   UP, LCP, APS

578.   ----. Gedanken von den
        Eigenschaften, Wirkungen
und Ursachen der Electricitaet,
nebst einer Beschreibung zwo
neuer electrischen Maschinen.
[24], 168p. 3pl. Leipzig, 1744.
Ms. notes (APS).
R; W313.   UP, LCP, APS

579.   ----. Die Stärke der elec-
        trischen Kraft des Wassers
in gläsernen Gefässen, welche
durch den Musschenbrökischen
Versuch bekannt geworden... [20],
164p. Leipzig, 1746.
R; W335.   APS

580.   ZALLINGER ZUM THURN, FRANZ
        SERAPHIM, 1743-1828. Ab-
handlung von der Elektricität des
in Tyrol gefundenen Turmalins...
[8], 59, [15]p. [Innsbruck], 1779.
R.   LCP

581.   ZETZELL, PER, 1724-1802.
        ...Consectaria electro-
medica...Upsaliae 1754 Octobr. 12.
In Caroli e Linné, Amoenitates
academicae...v.9. Erlangen, 1785.
(R).   CP

1801-1850

582.   ABBOT, JOEL, 1766-1826.
        An essay on the central in-
fluence of magnetism. 24p. illus.
Philadelphia, 1814.
Presented by the author.
Not in MRW.   APS

583.   AN ACT to incorporate the Sus-
        quehanna River & North & West
Branch telegraph co., passed April,
1849. 15p. n.p., 1849.
Not in MRW.   FI

584.   AIKIN, WILLIAM E   A   , M.D.
        An introductory lecture delivered
before the medical class of the University of Maryland, September, 1840.
32p. Baltimore, [1840].
History of electricity.
Not in MRW.   CP

AIRY, SIR GEORGE BIDDELL: see 660,
        749.

585.   ALDINI, GIOVANNI, 1762-1834.
        An account of the late im-
provements in galvanism, with a
series of curious and interesting
experiments...To which is added...
the author's experiments on the
body of a malefactor executed at
Newgate &c., &c. xi, 221p. 4pl.
London, 1803.
R; W644.   UP, LCP

50

586. ----. Essai théorique et experimental sur le galvanisme
avec une série d'expériences faite
en présence des commissaires de
l'Institut national de France, et
en divers amphithéatres anatomiques
de Londres. [4], x, 398p. 10pl.
Quarto. Paris, 1804.
Inscription by author to Joseph
Bonaparte.
Mp304, 366; R; W660.   FI, CP, APS

587. ----. Essai théorique et experimental sur le galvanisme,
... 2v. plates. Octavo. Paris,
1804.
Bookplate: Joseph Claude Anthelme
Recamier (LC).
R; W660.   LCP, CP

588. ----. Saggio di esperienze
sul galvanismo. [parts 1 &
2]. [2], 68p. Bologna, 1802.
Inscription by author: "Al citta-
dino consiliare Lambertenghi...
Sarà publicata la terza parte il
prossimo giugno colle tavole in
rame."
R.   APS

589. ALIBERT, JEAN LOUIS, 1766-
1837. Éloges historiques
composés pour la Société medicale
de Paris, suivis d'un discours
sur les rapports de la médecine
avec les sciences physiques et
morales. viii, 454p. Paris,
1806.
Mp240; R; (W632).   APS

ALLEN, ZACHARIAH: see 1252

590. ALMON, WILLIAM BRUCE.
...De galvanismo complec-
tens... [5], 63p. Edinburgh,
1809.
Inaugural dissertation.
Not in MRW.   LCP, APS

591. ALTRA RICADUTA del propa-
gatore ed ultimo rimedio
proposto alla sua guarigione
ossia. Ultima risposta contro
la difesa dei paragrandini, letta
all'Ateneo di Venezia, da un

socio di diverse accademie.
60p., [2] l. Milan, 1826.
Not in MRW.   APS

591a. AMORETTI, CARLO, 1741-1816.
Della raddomanzia ossia El-
lettromettria animale. Ricerche
fisiche e storiche. xviii, 476p.
6pl. Milan, 1808.
Mp401.   APS

592. AMPÈRE, ANDRÉ MARIE, 1775-
1836. Note sur l'action mutuelle d'un conducteur voltaique.
29p. 1pl. Paris, 1828.
Extrait des Ann. de Chim. et de
Phys., 1828.
R; W838bis.   CP

593. ---- Précis de la théorie
des phénomènes électro-dynamiques. Pour servir de supplément
à son Recueil d'observations électro-
dynamiques et au Manuel d'électricité
dynamique de M. Demonferrand. 67p.
1pl. Paris, 1824.
R.   APS

594. ----. Recueil d'observations
électro-dynamiques,...rélatifs
à l'action mutuelle de deux courans
électriques, à celle qui existe entre
un courant électrique et un aimant
ou le globe terrestre, et à celle de
deux aimants, l'un sur l'autre. 383
[i.e.361]p. 10pl. Paris, 1822.
Incomplete: pp.325-44 only, no plates.
Mp472; R; W784.   UP

595. ----. Théorie des phénomènes
électro-dynamiques, unique-
ment déduite de l'expérience. 226,
[2]p. 2pl. Paris, 1826.
Ms. errata addended.
Mp472; R.   FI, LCP, APS

see also: 760

ARELLA: see CARNEVALE-ARELLA

596. ASCHLUND, ARENT, 1797-1835.
Am compassets misviisning.
The variaition of the compass.
[8]p. 2 maps, table. [Copenhagen,
1831.]

Danish and English text.
Not in MRW.   APS

ASHBURNER, JOHN: see 917

597. ATLANTIC & OHIO TELEGRAPH CO.
      Report of the proceedings of
a meeting of the stockholders...
Philadelphia, July 19, 1849. 16p.
Philadelphia, 1849.
Not in MRW.   FI

598. ATLANTIC, LAKE & MISSISSIPPI
      TELEGRAPH. Exposure of the
schemes for nullifying the "O'Reilly
Contract" for extending the telegraph between the Atlantic, the
lakes & the Mississippi... 24,
28p. St. Louis, 1848.
Not in MRW.   FI

599. AUGUSTIN, FRIEDRICH LUDWIG,
      1776-1854. Versuch einer
vollständigen systematischen Geschichte der galvanischen Electricität und ihrer medicinischen Anwendung. xvi, 284p. 1pl. Berlin,
1803.
Mp383; R.   APS

600. ----. Vom Galvanismus und
      dessen medicinischer Anwen-
dung. iv, [2], 64p. front.
Berlin, 1801.
Ms. notes.
Mp572; R; W625.   APS

601. THE AURORA borealis, or an
      investigation into the causes,
and a solution offered for the
polar lights. By a mechanic. 69,
[3]p. 1pl. Canajahorie, N.Y.,
1835.
Not in MRW.   APS

602. BAADER, FRANZ VON, 1765-1841.
      Ueber den Blitz als Vater des
Lichts... 23p. [Munich, 1815].
R.   UP

603. BACCELLI, LIBERATO GIOVANNI,
      1772-1835. I fenomeni elet-
tro-magnetici a due leggi ridotti
con la lora cagione tolta dall'

opinione Symmeriana. Ragionamento
... 86p. Modena, 1821.
R.   APS

BACHE, ALEXANDER DALLAS: see 743,
      1243-46.

604. BACHHOFFNER, GEORGE HENRY,
      1810-1879. A popular trea-
tise on voltaic electricity &
electro-magnetism. 35p. London,
1838.
R; W928.   FI

BACHOUÉ: see LOSTALOT-BACHOUÉ

BACK, SIR GEORGE: see 659

605. BAGG, JOSEPH H   .
      On magnetism, or the doctrine
of equilibrium... x, 312p. illus.
Detroit, 1845.
Not in MRW.   FI

BAIN, ALEXANDER: see 724

606. BAIN, SIR WILLIAM, 1775-1853.
      Essay on the variation of
the compass, shewing how far it is
influenced by a change in the direction of the ship's head... [3],
140p. map. Edinburgh, 1817.
Mp457; R; W733.   LCP

607. BARLOW, JAMES.
      A new theory, accounting for
the dip of the magnetic needle,
being an analysis of terrestrial
magnetism, with a solution of the
lines of variation and no variation, and an explanation of the
nature of a magnet. xxvii, 183p.
plate, front. New York, 1835.
R; W891.   APS

608. BARLOW, PETER, 1776-1862.
      Essay on magnetic attrac-
tions and on the laws of terrestrial and electro-magnetism. 2d
ed. xii, 303p. plates. London,
1823.
Mp573; R; W765a.   FI

609. ----. Essay on magnetic at-

tractions...with an appendix, containing results of experiments made on ship board from latitude 61°S. to latitude 80°N. 2d ed. xii, 368p. 6pl. London, 1824.
Mp573; R.   FI

610. BARONIO, GIUSEPPE, ca. 1759-1811. Saggio di naturali osservazioni sulla elettricità voltiana colla descrizione d'una nuova macchina a corona di persone, & d'un piliere tutto vegetabile. 144p. Milan, 1806.
R.   APS

611. BECQUEREL, ANTOINE CÉSAR, 1788-1878. Elemente der Elektro-Chimie in ihrer Anwendung auf die Naturwissenschaften und die Künste...aus dem Französischen. xvi, 488p. 3pl. Erfurt, 1845.
R.   UP

612. ----. Éléments d'électrochimie appliquée aux sciences naturelles et aux arts. vi, 419p. 3pl. Paris, 1843.
Mp574; R.   FI, UP, APS

613. ----. Mémoire sur la phosphorescence produite par la lumière électrique, par Becquerel, Biot et Edmond Becquerel. [215]-241p. [Paris, 1839].
Reprint from Archives du muséum d'histoire naturelle, v.1.
From library of E. Newton Harvey.
R.   APS

614. ----. Traité complet du magnétisme. [3], cxi, 547p. 20pl. Paris, 1846.
Reprint of v.7 of author's Traité expérimentale de l'électricité et du magnétisme. (q.v.)
W1093.   APS

615. ----. Traité de physique considérée dans ses rapports avec la chimie et les sciences naturelles. 2v. 6pl. Paris, 1842-44.
Mp574; R.   APS

616. ----. Traité expérimental de l'électricité et du magné-

tisme, et de leurs rapports avec les phénomènes naturels;... 7v. in 9 (v.9 is Atlas). Paris, 1834-40.
Mp271 etc.; R; W882.   FI, UP, LCP (part), APS

BECQUEREL, EDMOND: see 613

617. BEEK, ALBERT VAN, 1787-1856. De l'influence que la fer des vaisseaux exerce sur la boussole ... 71p. plate. Paris, 1826.
Not in MRW.   FI, LCP

618. BEEZ, MARTIN. Einige Worte über die Erscheinungen der Electricität überhaupt. 56p. Würzburg, 1824.
Inaugural dissertation.
Not in MRW.   CP

619. BELLI, GIUSEPPE, 1791-1860. Sulla dispersione delle due elettricità e sui residui delle scariche delle bocce di Leida. Sperienze. 27p. 1pl. [Milan, 1837].
Memoria seconda & Memoria terza.
Inscription to Luigi Palmieri from the author.
R; W916.   APS

620. BELTRAMI, PAOLO. Buoni effetti dei paragrandini dell'anno 1825, e spiegazione del modo con cui questi semplici stromenti paralizzano le nubi temporalesche da impedire la formazione della grandine. Con appendice... 143p. Milan, 1826.
R.   APS

621. ----. Nuova scoperta importantissima comprovata dai più felici esperimenti per preservare le campagne dalla grandine devastatrice ed innaffiarle invece con pioggia ristoratrice. Presentata per la prima volta all'Italia... 2d ed. 27p. plate. Milan, 1823.
Mp389; R.   APS

622. ----. Riposta...alle critiche osservazioni del signor professore assistente Majocchi, con appendice di altra nuova probabile

scoperta... 26p. Milan, 1823
R.   APS

623. BERZELIUS, JÖNS JAKOB,
     friherre, 1779-1848. Af-
handling om galvanismen. [6],
145p. 1pl. Stockholm, 1802.
Presented by the author, 1829.
R.   APS

624. ----. An attempt to estab-
     lish a pure scientific system
of mineralogy, by the application
of the electro-chemical theory and
the chemical proportions...Tr. from
the Swedish original by John Black.
138p. London, 1814.
Presented by John Vaughan 1829.
R; W721.   UP, FI, APS

625. ----. Essai sur la théorie
     des proportions chimiques
et sur l'influence chimique de
l'électricité...tr. du suédois.
xvi, 190, 120, [2]p. Paris, 1819.
Presented by John Vaughan, 1840.
R; W755.   APS, FI, UP, CP

626. ----. Lärbok i kemien. 5v.
     plates. Stockholm, 1817-28.
V.1&2 are 2d ed.
Presented by author.
Not in MRW.   APS

627. ----. Lehrbuch der Chemie.
     5th ed. 5v. illus. plates.
Dresden & Leipzig, 1843-48.
Presented by Mrs. George F. Barker.
Mp370; R.   APS

628. ----. Traité de chimie...
     tr. par A.J.L. Jourdan...
[v.2-8 tr. par Me. Esslinger]. 8v.
plates. Paris, 1829-33.
Not in MRW.   APS

629. BEUDANT, FRANÇOIS SULPICE,
    1787-1852. Essai d'un cours
élémentaire et général des sciences physiques. Partie physique.
5th ed. xiv, 676p. 13pl. Paris,
1833.
Not in MRW.   APS

630. BEYER,   d. 1819.

Aux amateurs de physique,
sur l'utilité des paratonnerres.
116p. 3pl. Paris, 1809.
Mp389.   UP, APS

631. [BIANCHI, TOMMASO].
     Della vita del conte Ales-
sandro Volta. 138, [1]p. pl.,
port. Como, 1829.
R; W853.   APS

632. [BIOT, JEAN BAPTISTE], 1774-
     1862. Electricity. 191p.
2pl. n.p., n.d.
Translation of the chapters on
electricity in his Précis élémentaire de physique.
Inscription: R. M. Patterson Rec'd
from the Author Oct. 18, 1825.
Not in MRW.   APS

633. ----. Précis élémentaire de
     physique expérimentale. 2v.
plates. Paris, 1817.
APS has v.2 only.
(R); W809.   APS

634. ----. Précis élémentaire
     ...2d ed. 2v. plates.
Paris, 1821.
Inscription: "Doct. Patterson from
his friend W. E. Hanes(?) 1822."
(R).   APS

635. ----. Précis elementaire...
     3rd. ed. 2v. plates. Paris,
1824.
R; W809.   APS

see also: 265, 613, 725

636. BIRCH, JOHN, 1745-1815.
     An essay on the medical ap-
plication of electricity. iv, 57p.
London, 1802.
(R); W633bis.   CP

637. BIRD, GOLDING, 1815-1854.
     Lectures on electricity and
galvanism, in their physiological
and therapeutical relations, delivered at the Royal college of
physicians, rev. & extended. [2],
xii, 212p. illus. London, 1849.
R; W1153.   LCP, CP

54

638. BOMBAY GOVERNMENT observa-
tory. Observations made at
the magnetical and meteorological
observatory at Bombay, April-
December, 1845. 8, c, 276, 73p.
Bombay, 1846.
Presented by the Royal Society.
(R).   APS

639. BOMPASS, CHARLES CARPENTER.
An essay on the nature of
heat, light, and electricity. x,
266p. London, 1817.
Mp199; R; W735.   FI, LCP, APS

640. BOSTOCK, JOHN, 1773-1846.
An account of the history &
present state of galvanism. [2],
164p. 2pl. London, 1818.
Mp414, 443; R; W743.   FI, LCP,
                                   APS

641. BRANDELY, A
Operationen, manipulationen
und geräthschäften der electro-
chemie...aus dem französischen
bearbeitet von Friedrich Harzer.
179p. plates. Weimar, 1849.
R.   FI

642. BREMNER, JAMES.
Mystery of magnetism fully
discovered by experiments intui-
tively evident which admit of no
questions. 105p. London, 1825.
W817.   FI

BRESSY, JOSEPH: see 87-88

643. BREWSTER, SIR DAVID, 1781-
1868. A treatise on magnetism, forming the article under
that head in the seventh ed. of
the Encyclopaedia Britannica.
viii, 365p. illus., map. Edinburgh, 1837.
R.   LCP

see also: 923

644. BREWSTER, GEORGE, 1800-1865.
New philosophy of matter,
showing the identity of all the
imponderables and the influence

which electricity exerts over matter in producing all chemical changes
and all motion. 216p. Boston, 1843.
Not in MRW.   UP

645. BRITISH ASSOCIATION for the
advancement of science. Proceedings connected with the magnetical and meteorological conference,
held at Cambridge in June 1845.
78p. London, 1845.
Inscription by Lt. Maury.
R.   APS

646. BROSSAUD, ÉMILE.
Essai sur les différentes
espèces d'électricité, appliquées
au traitement des affections ner-
veuses et rheumatismales. 31p.
Paris, 1828.
Inaugural dissertation.
Not in MRW.   CP

647. BROUN, JOHN ALLAN, 1817-1879.
Report to General Sir Thomas
Makdougall Brisbane, Bart., on the
completion of the publication, in
the Transactions of the Royal society of Edinburgh, of the observations made in his observatory at
Makerstoun. 17p. Edinburgh, 1850.
R.   APS

see also: 858

648. BROWN, THOMAS, of Troy.
The ethereal physician; or,
Medical electricity revived; its
pretensions fairly and candidly
considered and examined, and its
efficacy proved, in the prevention
and cure of a great variety of
diseases; with the details of up-
wards of sixty cures...with some
observations on the nature of the
electric fluid, and hints concerning the best mode of applying it
for medical purposes. No. I...To
which is added a brief account of
its medical practice by Jesse
Everett. vi, [2], 64, 10, [2]p.
front. Albany, 1817.
Not in MRW.   APS

649. ----. The ethereal physician; or, The medical powers of electricity demonstrated...No. II. By Thomas Brown...and Jesse Everett. [83]-152, [2]p. Albany, 1823.
Not in MRW.   CP

BRUSSELS. OBSERVATOIRE ROYALE: see 913

650. BUFFON, GEORGE LOUIS LECLERC, conte de, 1707-1788. Traité de l'aimant et de ses usages. pp.293-452 v.14; 1-429 v.15; 1-60 v.16. plates. Paris, 1802.
In his Histoire naturelle, générale et particulière...Nouvelle éd... redigé par C. S. Sonnini.
(R).   APS

651. BYWATER, JOHN.
An essay on light and vision ...To which are added, some original remarks and experiments on the magnetic phenomena, intimately connected with practical navigation. 89p. Liverpool, 1813.
Not in MRW.   CP

652. ----. An essay on the history, practice, & theory, of electricity. iii, 127p. 2pl. London, 1810.
Presented by the author, 1827.
R; W701.   APS

653. CARNEVALE-ARELLA, ANTONIO. Storia dell' elettricità. 2v. in 1. 230, 272p. Alessandria, 1839.
Mp296; R; W952.   APS

654. CARPUE, JOSEPH CONSTANTINE, 1764-1846. An introduction to electricity and galvanism; with cases, showing their effects in the cure of diseases. To which is added, a description of Mr. Cuthbertson's plate electrical machine...Being the substance of lectures delivered to his anatomical class. viii, 112p. 3pl. London, 1803.
Mp375; R; W646.   APS

655. CARRADORI, GIOACHINO, 1758-1818. Istoria del galvanismo in Italia ossia della contesa fra Volta e Galvani... 72p. Florence, 1817.
R.   APS

CAUSTIC, CHRISTOPHER, pseud.: see FESSENDEN

656. CHANNING, WILLIAM FRANCIS, b. 1820. Notes on the medical application of electricity. 199p. illus. Boston, 1849.
Mp584.   FI, CP

657. CHAPPE, IGNACE URBAIN JEAN, 1760-1828. Histoire de la télégraphie. 2v. plates. Paris, 1824.
R; W810.   FI

658. CHEVALIER, CHARLES LOUIS, 1804-1859. Nouvelles instructions sur l'usage du daguerréotype. Description d'un nouveau photograph, et d'un appareil très simple destiné à la réproduction des épreuves au moyen de la galvanoplastie... 78, [1]p. 1pl. Paris, 1841.
Not in MRW.   APS

659. CHRISTIE, SAMUEL HUNTER, 1784-1865. Discussion of the magnetical observations made by Captain Back, R.N., during his late Arctic expedition. pp.377-415. tables. London, 1836.
From Phil. Trans., part II for 1836.
R; W2703.   APS

660. ----. Report upon a letter addressed by M. le Baron de Humboldt to his Royal Highness the President of the Royal Society, and communicated by His Royal Highness to the Council. [By] S. Hunter Christie [and] G. B. Airy. pp.418-428. London, 1836.
Photocopied from Royal Society of London Proceedings, v.3.
Mp335; R;(W2720).   APS

661. COAD, PATRICK.

An illustration of Coad's patent graduated galvanic battery, and his patent insulated poles, for medical purposes, with certificates of the most astonishing cures, effected by means of this highly useful apparatus. References to numerous eminent physicians, the outlines of his lectures on galvanism, &c. &c. with directions respecting the sale of his patent rights and privileges. 24p. Philadelphia, 1844.
Not in MRW.   CP

662. COOPER, C   CAMPBELL.
An attempt to unite the different theories concerning light, heat, electricity, galvanism and magnetism. pt. 1; identity of caloric and electricity. 20p. Philadelphia, 1842.
Presented by the author, 1842. (Mp587).   APS

663. ----. Identities of heat, & light, caloric & electricity. 96p. Philadelphia, 1848.
Mp587; R.   UP, CP

664. COUDRET, J   F   .
Récherches médico-physiologiques sur l'électricité animale, suivies d'observations et de considérations pratiques sur le procédé médical de la neutralisation électrique directe... viii, 496, 3p. 3pl. Paris, 1837.
R.   CP

COURTENAY, EDWARD HENRY: see 1244

665. CRUSELL, GUSTAV SAMUEL, 1810-1858. Über den Galvanismus als chemisches Heilmittel gegen örtliche Krankheiten. Mit einem Schreiben von M. Markus. 68p. 1pl. St. Petersburg, 1841.
Erster Zusatz. pp. [69]-84. St. Petersburg, 1842.
R.   APS

CUMMING, JAMES: see 683

666. CUNNINGHAM, PETER MILLER, 1789-1864. On the motions of the earth & heavenly bodies as explainable by electro-magnetic attraction & repulsion & on the conception, growth & decay of man & cause & treatment of his diseases as referable to galvanic action. 281p. London, 1834.
R; W883.   FI

667. CUTHBERTSON, JOHN.
Practical electricity & galvanism; containing a series of experiments calculated for the use of those who are desirous of becoming acquainted with that branch of science. xx, 271p. plates. London, 1807.
Mp589; R; W681.   FI, LCP

668. ----. Practical electricity and galvanism;... 2d ed. xxiii, 294, 8p. 9pl. London, 1821.
R; W681a.   APS

669. DAVENPORT, THOMAS, 1802-1851. Exhibition of models of Davenport's electro-magnetic machinery, is now open to the public, day & evening, at Masonic hall, Chestnut st... 4p. n.p. [183-].
Not in MRW.   APS

see also: 750, 959

670. DAVIES, JOHN.
Remarkable case of the effects of lightning on the human body; with general observations on the nature and phenomena of lightning, adapted for general readers. 24p. London, 1839.
Not in MRW.   CP

671. DAVIS, DANIEL, JR.
Davis' manual of magnetism: including also electromagnetism, magneto-electricity; with a description of the electrotype process. viii, 218p. illus. Boston, 1842.
(R); W1012.   FI

672. ----. Manual of magnetism,
including galvanism, magnetism, electro-magnetism, electro-
dynamics, magneto-electricity, and
thermo-electricity. 2d ed. viii,
322p. illus., front. Boston, 1847.
(W1012a).   UP, APS

673. ----. Manual of magnetism
... 2d ed. viii, 322p.
Boston, 1848.
W1012a.   FI

674. [DAVIS, DANIEL, Jr.]
The medical application of
electricity; with descriptions of
apparatus, and instructions for its
use. 2d ed. 26p. illus. Boston,
1847.
Mp590.   APS

675. DAVY, SIR HUMPHRY, 1778-1829.
Collected works; ed. by his
brother, John Davy. 9v. plates,
port. London, 1839-40.
Mp347; R; W956.   APS

676. ----. Élémens de philosophie
chimique...tr. de l'anglais,
avec des additions, par. J.-B. Van
Mons. 2v. plates. Paris, 1826.
Presented by John Vaughan, 1838.
Not in MRW.   APS

677. ----. Elements of chemical philosophy. Part I.
Vol. I [all published]. xii,
296p. 10pl. Philadelphia, 1812.
Mp341; R; W710.   APS

678. ----. Six discourses delivered before the Royal
society. v.p. London, 1827.
R; W829.   APS

679. ----. Syllabus of a course
of lectures on chemistry, delivered at the Royal institution...
[3], 91p. London, 1802.
W634.   UP

see also: 887

680. DE KRAMER, ANTONIO, 1806-

1853. Nuovo apparato rota-
torio elettro-magnetico messo in
moto dal magnetismo terrestre.
Nota communicata alla Biblioteca
italiana. 7p. plate. Milan, 1838.
R.   APS

DELAUNAY, CLAUDE VEAU: see VEAU DE
LAUNAY

681. DELESSE, ACHILLE ERNEST OSCAR
JOSEPH, 1817-1881. Sur le
pouvoir magnétique des minéraux et
des roches. 59p. Paris, [1848].
From Annales des mines, 1848.
Presented by the author.
R.   APS

682. ----. Sur le pouvoir magnétique des roches (cont.)
22p. Paris, [1849].
From Annales des mines, 1849.
R.   APS

683. DEMONFERRAND, JEAN BAPTISTE
FIRMIN, 1795-1844. Manual
of electro-dynamics, chiefly tr.
from the Manuel d'électricité dyna-
mique...of J. F. Demonferrand.
With notes and additions,...by
James Cumming. viii, 291p. plates.
Cambridge [Engl.], 1827.
R; W827.   UP

684. ----. Manuel d'électricité
dynamique; ou, Traité sur
l'action mutuelle des conducteurs
électriques et des aimans, et sur
une nouvelle théorie du magnétisme;
pour faire suite à tous les Traités
de physique élémentaire. 216, [2]p.
5pl. Paris, 1823.
R; W797.   LCP, APS

685. DERMOGINE, EPUGISPE.
Comenti...alla difesa di Angelo Bellani della lettera supposta
del Signor Conte Volta al Signor
Marzari...con osservazioni di X.Z.
38 [2]p. Milan, 1823.
Not in MRW.   APS

686. DESPRETZ, CÉSAR MANSUÈTE,
1792-1863. Traité élémentaire
de physique... 4th ed. vi, 918p.

58

plates. Paris, 1836.
W903.   UP

687. [DOLLOND, PETER], 1730-1820.
      Directions for using the
electrical machine, made by P. and
J. Dollond, opticians to His
Majesty. 28p. 1pl. London, 1802.
(Mp214); (R); (W405).   APS

688. DONOVAN, MICHAEL, 1790-
      1876. Essay on the origin,
progress & present state of galvanism; containing...a new hypo-
thesis. 390p. plates. Dublin,
1816.
R; W730.   FI, LCP

689. ----. Essay on the use and
      employment of electricity,
electro-magnetism, and magneto-
electricity in the cure of diseases, collected from various
sources. 80p. Dublin, 1848.
Inscription by the author to Dr.
Wood.
Not in MRW.   CP

690. DOYLE, GEORGE S
      Mysteries philosophically ex-
plained: Newton's theory of attraction confuted, animal magnetism,
the magnetic needle, & gravitation
demonstrably accounted for on
philosophic principles. By George
S. Doyle & S. Albert Whitney. 30p.
Boston, 1843.
Not in MRW.   FI

690a. DRAPER, JOHN WILLIAM, 1811-
      1882. A treatise on the
forces which produce the organiza-
tion of plants. With an appendix,
containing several memoirs on cap-
illary attraction, electricity, and
the chemical action of light. xi,
108, 216p. 3pl., front. New York,
1844.
Also 1845 ed. xvi, 216p. 3pl.,
front. New York, 1845.
Not in MRW.   APS

691. DUBOIS-REYMOND, EMIL HEINRICH,
      1818-1896. Quae apud veteres

de piscibus electricis exstant ar-
gumenta... 30, [2]p. Berlin,
[1843].
Inaugural dissertation.
R.   APS

692. ----. Untersuchungen über
      thierischen Elektricität.
2v. in 3. 11pl. Berlin, 1848-60.
R.   CP, FI(v.1)

693. DUPREZ, FRANÇOIS JOSEPH FER-
      DINAND, 1807-1884. Memoire
en réponse à la question suivante:
On demande un examen approfondi de
l'état de nos connaissances sur
l'électricité de l'air, et des
moyens employés jusqu'à ce jour
pour apprécier les phénomènes électriques qui se passent dans l'atmosphère. 134p. [Brussels, 1843].
R.   FI

694. EDWARDS, WILLIAM FRÉDÉRIC,
      1777-1842. De l'influence
des agens physiques sur la vie.
xvi, 654, [2]p. 1pl. Paris, 1824.
R.   APS, UP

695. ----. On the influence of
      physical agents on life...Tr.
from the French by Dr. Hodgkin and
Dr. Fisher. To which are added in
the appendix, some observations on
electricity:...and some notes to the
work of Dr. E. [Ed. by Dr. Hodgkin.]
488p. London, 1832.
Not in MRW.   LCP

696. ----. On the influence of
      physical agents...To which
are added, in the Appendix, some
observations on electricity, by Dr.
Edwards, M. Pouillet, and Luke Howard...and some notes to the work of
Dr. Edwards. 240p. Philadelphia,
1838.
Not in MRW.   UP, CP

697. ELECTRICITY.
      64p. illus. n.p., n.d.[18--].
Textbook, interleaved with blank
pp., ms. notes.
No t.p.
Not in MRW.   UP

698. ELIAS, PIETER, 1809-1878.
    Beschrijving eener nieuwe
machine ter aanwending van het
electromagnetismus als beweeg-
kracht. 40p. 1pl. Haarlem, 1842.
Tipped in: Magneto electrische
machine (Broadside).
R; W1015.   APS

see also: 1289

699. ELICE, FERDINANDO, b. 1786.
    Instruzione sui parafulmini;
lettera indirizzata al Pittore
Costantino Dentone. 2d ed. 31p.
Genoa, 1841.
Presented by author 1841.
R.   APS

700. ----. Notizie elettriche.
    [Letter to Luigi Foppiani].
12, [1]p. Genoa, [1844?].
Presented by the author.
R.   APS

701. ----. Osservazioni sull'
    instruzione de' parafulmini
approvata dalla R. Accademia delle
scienze di Parigi il di 23 Aprili
1823 e pubblicata nel 1824. 8p.
Genoa, 1826.
Presented by the author.
R.   APS

EVERETT, JESSE: see 648-9

702. EXLEY, THOMAS, 1775-1855.
    Principles of natural philo-
sophy: or, A new theory of physics,
founded on gravitation, and applied in explaining the general
properties of matter, the phenomena
of chemistry, electricity, galvanism, magnetism, & electro-magnetism.
xxxii, 478p. 4pl. London, 1829.
R; W848.   APS

703. EYDAM, IMMANUEL, 1802-1847.
    Die Erscheinungen der Elek-
tricität und des Magnetismus in
ihrer Verbindung mit einander.
Nach den neuesten Entdeckungen im
Gebiete des Elektro-Magnetismus
und der Induktions-Elektricität
... xx, 360p. 3pl. Weimar, 1843.
Mp601; R; W1039.   APS

FABRÉ-PALAPRAT, BERNARD-RAYMOND:
    see 818

704. FANSHER, SYLVANUS.
    A concise treatise on elec-
tricity; with directions for constructing and applying lightning
rods, used for the purpose of
shielding houses, ships, etc.,
against the dangerous effects of
electrical fluid from the thunder-
cloud. 36p. illus., plates. New
Haven, 1830.
Not in MRW.   HSP

705. FARADAY, MICHAEL, 1791-1867.
    An answer to Dr. Hare's let-
ter on certain theoretical opinions
... [108]-120p. n.p., n.d.
From American journal of science,
1840.
Inscribed by the author to APS.
Not in MRW.   APS

706. ----. Chemical manipulation;
    being instructions to stu-
dents in chemistry... vii, 656p.
illus. London, 1827.
APS has 3rd ed. London, 1842 also.
(R).   APS

707. ----. Experimental researches
    in electricity. 3v. 17pl.
London, Quaritch, 1839-55.
Reprinted from Phil. Trans., 1831-52.
Autograph: Elihu Thomson, 1885 (APS).
Presented by Penrose R. Hoopes, 1968.
M; R; W959 & 959a.   APS, FI

708. ----. Experimental researches
    ... 3v. 17pl. London, Taylor,
1839-55.
Reprinted from Phil. Trans., 1831-52.
Signed: from the author (FI)
Presented by Charles N. Bancker (APS).
M; R; W959 & 959a.   LCP, CP,
                      FI(1-21),
                      APS(v.1, 1839)

709. ----. Experimental researches
    in electricity. (18th series)
No. 25. On the electricity evolved
by the friction of water & steam
against other bodies. pp.17-32.
1pl. London, 1843.
From Phil. Trans. Part I for 1843.

60

Inscribed by the author.
W2801.   APS

710. ----.   Experimental researches
      in electricity...reprinted
from the Phil. Trans. of 1831-43.
With other electrical papers from
the Qtrly. J. of Sci. & Phil. Mag.
2v. plates. London, 1844-49.
V.1 is 2d ed.
W959a.   UP

711. ----.   On the general mag-
      netic relations and charac-
ters of the metals: additional
facts. pp.161-3. n.p., 1839.
From the London and Edinburgh
philosophical magazine...Mar.
1839.
Not in MRW.   APS

712. ----.   On static electri-
      cal inductive action. Let-
ter to R. Phillips 4 Feb. 1843.
4p. n.p., n.d.
Not in MRW.   APS

see also: 915

713. FARDELY, WILLIAM.
      Die Galvanoplastik, oder
Praktische Anleitung, Metalle aus
ihren Auflösung nach den neuesten
und verbesserten Verfahrungsarten,
vermittelst der galvanischen Electricität zu reduciren... 47p. 1pl.
Mannheim, 1842.
R.   CP

714. FARRAR, JOHN, 1779-1853, comp.
      Elements of electricity, mag-
netism, and electro-magnetism,...
being the second part of a course
of natural philosophy, compiled for
the use of the students of the university at Cambridge, New England.
vii, 395p. 6pl. Cambridge, [Mass.],
1826.
Ms. notes.
Mp292 etc.; R.   FI, APS

715. ----.   Elements of electri-
      city, magnetism, and electro-
dynamics... 376p. plates. Boston,

1839.
Mp601; R.   FI

716. ----.   Elements of electri-
      city, magnetism, & electro-
dynamics,...for the use of the
students of Harvard University; being
the second part of a course of natural philosophy by John Farrar &
the first part of a new course of
physics, by Joseph Lovering. vii,
336p. 6pl. Boston, 1842.
Mp601; R.   FI

717. FECHNER, GUSTAV THEODOR, 1801-
      1887. Elementar-lehrbuch des
Elektromagnetismus nebst Beschrei-
bung der hauptsächlichen elektro-
magnetischen Apparate. 157p. plates.
Leipzig, 1830.
R.   FI

718. ----.   Reportorium der Ex-
      perimentalphysik... 3v.
10pl. Leipzig, 1832.
V.1 lacking.
W865.   LCP

719. [FESSENDEN, THOMAS GREEN],
      1771-1837. A poetical peti-
tion against tractorising trumpery,
and the Perkinistic Institution.
In four cantos. Most respectfully
addressed to the Royal college of
physicians, by Christopher Caustic.
92p. London, 1803.
W647.   CP

720. ----.   Terrible tractoration!!
      A poetical petition against
galvanising trumpery, and the Perkinistic institution... 2d ed.
xxxi, 186p. front. (colored).
W647a.   CP

721. ----.   Terrible tractoration!!
      ... 1st Amer. from 2d London
ed. xxxv, 192p. front. New York,
1804.
Presented by Jonathan B. Smith (APS).
Mp328;(W647a).   CP, APS

722. ----.   The modern philosopher;
or, Terrible tractoration!... 2d
Amer. ed. xxxii, 270p. front.

Philadelphia, 1806.
Not in MRW.   CP

723. [----]. Terrible tractora-
      tion, and other poems, by
Christopher Caustic. 3rd Amer.
ed. viii, 264p. front. Boston,
1836.
Not in MRW.   CP

724. FINLAISON, JOHN, 1783-1860.
      Account of some remarkable
applications of the electric fluid
to the useful arts by Alexander
Bain with a vindication of his
claim to be the first inventor of
the electro-magnetic printing telegraph & also of the electro-magnetic clock. 127p. plate. London,
1843.
W1040.   FI

725. FISCHER, ERNST GOTTFRIED,
      1754-1831. Physique méca-
nique...tr. de l'allemand.
Avec des notes de M. Biot... 2.éd.
xiv, 484p. 8pl. Paris, 1813.
Autograph: Robert Maskell Patterson.
Also 3rd & 4th eds. Paris, 1819-
1830.
Not in MRW.   APS

726. FISCHER, NICOLAUS WOLFGANG,
      1782-1850. Verhältniss der
chemischen Verwandschaft zur galvanischen Elektricität in Versuchen dargestellt. 238p. Berlin,
1830.
R.   FI

727. FISHER, GEORGE THOMAS, 1822-
      1847. Practical treatise on
medical electricity, containing a
historical sketch of frictional and
voltaic electricity, as applied to
medicine: with plain instructions
for the use of electric, galvanic,
& electro-magnetic instruments: and
embracing an account of the most
recent researches of Matteucci.
73p. illus. London, 1845.
Mp467.   CP

728. FORBES, JAMES DAVID, 1809-

1868. Account of some additional experiments on terrestrial
magnetism, made in different parts
of Europe in 1837. [2], 27-36p.
tables. Edinburgh, 1840.
From the Trans. of the Royal society of Edinburgh, v.XV, pt. 1.
Inscribed by the author.
R; W2723a.   APS

729. ----. Account of some ex-
      periments made in different
parts of Europe, on terrestrial
magnetic intensity, particularly
with reference to the effect of
height. [2], 29p. plate. Edinburgh, 1837.
From the Trans. of the Royal society of Edinburgh, v.XIV.
Inscribed: from the author.
R; W2723.   APS

FORBES, ROBERT BENJAMIN: see 784

730. [FOX, ROBERT WERE], 1789-
      1877. Instructions for using
Mr. Fox's instrument for determining
the magnetic inclination and inten-
sity. 15p. illus. London, 1842.
Inscription: from R. W. Fox -
March 8, 1843.
Not in MRW.   APS

731. FRIESE, ROBERT.
      Theoria galvanismi. [with
vita.] 80, [2]p. Bonn, 1842.
Inaugural dissertation.
R; W1016.   CP

732. FRORIEP, ROBERT, 1804-1861.
      Beobachtungen über die Heil-
wirkung der Electricität, bei Anwendung des magnetoelectrischen
Apparates...Erstes Heft. Die rheu-
matisches Schwiele. xxxvi, 292p.
illus. Weimar, 1843.
Mp604; R.   CP

733. ----. On the therapeutic ap-
      plication of electro-magnetism
in the treatment of rheumatic and
paralytic affections...Tr. from the
German by Richard Moore Lawrance.
205p. London, 1850.
R; W1175.   CP, UP

734. GALE, T[HADDEUS?].
Electricity or ethereal fire,
considered 1st. Naturally, as the
agent of animal and vegetable life:
2nd. Astronomically, or as the agent
of gravitation and motion: 3rd.
Medically, or its artificial use
in diseases, comprehending both the
theory & practice of medical electricity... 276, [3]p. Troy, 1802.
Mp364; W636.   FI, CP, APS

735. GAUSS, KARL FRIEDRICH, 1777-
1855. Intensitas vis mag-
neticae terrestris ad mensuram ab-
solutam revocata. 44p. Göttingen,
1833.
R; W867.   APS

736. ----. Resultate aus den Beobachtungen des magnetischen
Vereins im Jahre 1836. Heraus-
gegeben von Carl Friedrich Gauss
und Wilhelm Weber. 103, [21]p.
10pl. Göttingen, 1837.
Mp605; R; W920.   LCP, APS

737. GAVARRET, LOUIS DOMINIQUE
JULES, 1809-1890. Lois gen-
erales de l'électricité dynamique,
analyse et discussion des principaux phénomènes physiologiques et
pathologiques qui s'y rapportent.
151p. Paris, 1843.
Inaugural dissertation.
R; W1042.   CP

738. GAY-LUSSAC, JOSEPH LOUIS,
1778-1850. Instruction sur
les paratonnerres, adoptée par
l'Académie des Sciences, le 23 juin
1823. [Signé Poisson et al.] 51p.
2pl. Paris, 1824.
Mp648; R.   APS, FI

739. GEMMINGER, MAX, 1820-1887.
Elektrische Organ von Mor-
myrus und Schwanzskelet von Eryx.
15p. 1pl. Munich, 1847.
R.   CP

740. GHERARDI, SILVESTRO, 1802-
1879. Osservazioni intorno
ad un articolo del chiarissimo Sig.
Prof. G. Grimelli sulla collezione
Galvaniana...e nuove notizie sul
Galvani illustrazioni e difese
delle opere sue. Discorso...con
appendice sopra un edizione di
opere del cel. P. Beccaria pochis-
simo conosciuta in cui si fa men-
zione di un opuscolo del Galvani.
58p. Bologna, 1842.
R.   APS

741. GIANNI, FRANCESCO, 1759-1822.
Raccolta della poesie di
Francesco Gianni. 2v. in 1. Milan,
1807.
For poem on "L'elettricita" see
pt.2, pp.38-39.
Not in MRW.   APS

742. GILLISS, JAMES MELVIN, 1811-
1865. Magnetical & meteorological observations made at Washington...1838-42. xxviii, 648p.
21 charts. Washington, 1845.
Presented & inscribed by the author.
R; W1073.   FI, APS

743. GIRARD COLLEGE, Philadelphia.
Observations at the Magnetical and meteorological observatory...made under the direction of
A. D. Bache and with funds supplied
by members of the American philosophical society and by the Topo-
graphical bureau of the United
States, 1840-1845. 3v. 1v, 3212p.
1 vol. of plates. Washington, 1847.
R.   APS

744. GORHAM, JOHN, 1783-1829.
The elements of chemical
science. 2v. plates. Boston,
1819-20.
APS has v.1 only.
Presented by the author.
Not in MRW.   APS

745. GREAT BRITAIN - Ordnance &
admiralty, Dept. of. Observations on days of unusual magnetic
disturbance; made at the British
colonial magnetic observatories,
1840-44...Under the superintendence
of Lieut.-Colonel Edward Sabine.
v.1, pt. 1-2. London, 1843-51.
Presented by the British Government
R; W1050.   FI, APS

746. GREEN. GEORGE, 1793-1841.
Essay on the application of mathematical analysis to the theories of electricity & magnetism. 72p. Nottingham, 1828.
Mp609; R; W840.   FI

747. GREEN, JACOB, 1790-1841.
Electro-magnetism: being an arrangement of the principal facts hitherto discovered in that science ... 216p. 4pl. Philadelphia, 1827.
Presented by the author, 1827.
Mp609; R; W831.   UP, HSP, APS

748. [----]. An epitome of electricity & galvanism, by two gentlemen of Philadelphia [Jacob Green and Ebenezer Hazard]. [8], xlviii, [3], 159, [8]p. front.
Philadelphia, 1809.
From library of Samuel Pennypacker (APS).
Mp609; W697.   FI, UP, HSP, CP
APS, LCP

749. GREENWICH. ROYAL Observatory. Magnetical and meteorological observations made in the years 1840-47...under the direction of George Biddell Airy... 7v.
London, 1843-49.
R.   APS

GREGORY, WILLIAM: see 916, 918

750. GRIGLIETTA, C   .
A brief essay or informal lecture on electro-magnetism, with a full description of models of Davenport's machines, as now exhibited...with an extract from the American journal of science, by Professor Silliman... 24p. Philadelphia, 1838.
Not in MRW.   LCP, CP

751. GRIMELLI, GEMINIANO.
Memoria...sul galvanismo premiata dall'accademia delle scienze dell'Istituto di Bologna...
193p. Bologna, 1849.
R.   APS

see also: 740

752. GRIMSTON, HENRY.
Apology for believing in the metallic tractors, with some account of the Perkinean institution for the benefit of the poor. 2d ed. To which is now added a postscript, containing some recent cases, particularly in disorders of the eye. 53p. London, 1805.
Not in MRW.   CP

753. GROTHUSS, THEODOR, freiherr von, 1785-1822. Über die chemische Wirksamkeit des Lichts und der Elektricität; besonders über einen merkwürdigen neuen Gegensatz dieser Wirksamkeit, den das Licht auf gewisse Substanzen äussert, je nachdem es entweder aus nichtoxidirenden Körpern oder aus der atmosphärischen Luft unmittelbar in dieselben und aus letzteren in jene eindringt. 76p.
[Riga?, 1818].
R.   UP

754. GRUND, FRANCIS JOSEPH, 1805-1863. Elements of natural philosophy, with practical exercises, for the use of schools. xii, 278p. illus. Boston, 1832.
Not in MRW.   APS

755. GUSSEROW, CARL AUGUST.
De electricarum chemiarumque organismi virium ratione atque efficacia. 41p. Berlin, 1832.
Inaugural dissertation.
Not in MRW.   CP

756. HALL, RICHARD WILLMOTT, respondent. ...On the use of electricity in medicine... 35p.
Philadelphia, 1806.
Inaugural dissertation.
Not in MRW.   UP, CP, APS

757. HALSE, WILLIAM HOOPER.
On medical galvanism. 25p.
illus. London, [1844].
Not in MRW.   CP

758. HANSTEEN, CHRISTOPHE, 1784-1873. De mutationibus, quas subit momentum virgae magneticae

64

partim ob temporis, partim ob tem-
peraturae mutationes. 43p. tables.
Christiania [Oslo], 1842.
Presented by the Royal University,
Christiania, 1843.
(R?).   APS

see also: 942

759. HARE, ROBERT, 1781-1858.
        Animadversions, on the re-
view of his theory of galvanism, by
Dr. Patterson. Published in the
first number of the Philadelphia
medical and physical journal. 18p.
[Philadelphia?, 1820?].
W683.   FI, UP, AFS

760. ----. A brief account of the
        theory of Ampère. 2p. n.p.,
n.d.
Not in MRW.   CP

761. ----. A brief exposition of
        the science of mechanical
electricity... viii, 48p. illus.
Philadelphia, 1835.
W895.   FI, CP

762. ----. A brief exposition of
        the science of mechanical
electricity, or electricity proper;
subsidiary to the course of chemical instruction in the University
of Pennsylvania... vii, 56p. illus.
Philadelphia, 1840.
With his Compendium of the course
of chemical instruction... 4th
ed.
Inscription: To George B. Emerson,
Esq. with the author's regards.
(W895).   FI, UP, CP, APS

763. ----. Compendium of the
        course of chemical instruc-
tion in the medical department of
the University of Pennsylvania...
xix, 310p. illus. Philadelphia,
1828.
Contains his Brief exposition...
mechanical electricity, 1835.
Also 2d, 3rd, & 4th eds. Philadelphia, 1834-40.
Not in MRW.   CP

764. ----. An essay on the ques-

tion, whether there be two electrical fluids, according to Du Faye,
or one, according to Franklin.
10p. 1pl. n.p., [1823?].
No t.p.
R.   UP, APS

765. ----. Essays on electricity.
        Being a portion of [his] Com-
pendium of the course of chemical
instruction...published in advance
for the use of his students. 23p.
3pl. [Philadelphia, 1830].
Not in MRW.   CP

766. ----. Experimental observa-
        tions, and improvements in
apparatus and manipulation; with
theoretical suggestions respecting
the causes of tornadoes, falling
stars, and the aurora borealis.
60p. illus., plate, front. Philadelphia, 1836.
From APS Transactions, n.s. v.5.
Inscribed to Robert Walsh from the
author.
R.   APS, CP

767. ----. Exposition of the sci-
        ences of galvanism & electro-
magnetism abstracted from 5th ed.
of Turner's chemistry. 37, 11p.
illus. n.p., n.d.
No t.p.
Not in MRW.   FI, UP, CP

768. ----. A letter to Prof. Fara-
        day on certain theoretical
opinions. 11p. No t.p.
From American Journal of Science,
v.38, 1840.
R; W2766.   APS

769. ----. A memoir on some new
        modifications of galvanic ap-
paratus, with observations in sup-
port of his new theory of galvanism.
17p. 1pl. n.p., [1821?].
R; W769.   UP, CP, APS

770. ----. Minutes of the course
        of chemical instruction in
the medical department of the University of Pennsylvania. For the use
of his pupils. Reprinted, with corrections and additions. 96, 62, 69,

56p. plates. Philadelphia, 1825.
Contains: Lectures on electricity
& galvanism & an appendix on electricity.
Not in MRW.   UP, APS

771. ----. A new theory of gal-
     vanism, supported by some
experiments and observations made
by means of the calorimotor, a new
galvanic instrument. Also, a new
mode of decomposing potash extem-
poraneously. Read before the
Academy of natural sciences, Philadelphia. 17, [2]p. 1pl. Philadelphia, 1819.
Autograph: John Vaughan.
R.   FI, UP, APS

772. ----. Objections to the the-
     ories severally of Franklin,
Dufay and Ampere, with an effort to
explain electrical phenomena, by
statical, or undulatory polariza-
tion. 24p. [Philadelphia, 1847].
From Medical examiner, n.s. v.3,
"republished with corrections and
additions by the author."
Not in MRW.   CP

773. ----. Objections to the the-
     ories severally of Franklin,
Dufay and Ampère, with an effort to
explain electrical phenomena, by
statical, or undulatory, polariza-
tion. (An improved edition, Feb-
ruary 17, 1848). 22p. Philadelphia, 1848.
Not in MRW.   UP, APS

774. ----. Of galvanism, or vol-
     taic electricity. 80p. illus.
n.p., [1843?].
Paragraphs no. 300-589.
Not in MRW.   CP

775. ----. On electricity. 39p.
     n.p., n.d.
Not in MRW.   UP

776. [----]. On electricity, &
     Appendix to lectures on elec-
tricity, galvanism, &c. 40, 23p.
4pl. n.p., n.d.
Not in MRW.   FI, UP, CP

777. [----]. On the origin and
     progress of galvanism, or
voltaic electricity. 16p. n.p.,
n.d.
Not in MRW.   LCP, APS

778. [----]. On the origin & pro-
     gress... 24p. illus. plate.
n.p., n.d.
No t.p.
Not in MRW.   FI

779. ----. On the question whether
     the discordancy between the
characteristics of mechanical electricity, and the galvanic or voltaic
fluid, can arise from "the difference
of intensity and quantity." With
some observations in favour of the
existence of electro-motive power,
independently of chemical reaction,
but co-operating therewith. 3-11p.
[Philadelphia, 1836?].
R.   CP

780. ----. Second letter to Pro-
     fessor Faraday. 15p. n.p.,
1841.
Not in MRW.   CP, APS

781. ----. Strictures,...in re-
     ply to remarks on the calori-
motor, published at different times,
in the American medical recorder.
28p. 1pl. Philadelphia, 1820.
Autograph: A. Seybert (c.2 APS).
Not in MRW.   FI, UP, CP, APS

782. ----. Supplement to the
     minutes of the course of
chemical instruction... 69, 56,
[2]p. 4pl. Philadelphia, 1824.
Not in MRW.   CP

see also: 1252-55

783. HARRIS, WILLIAM SNOW, 1791-
     1867. On the nature of
thunderstorms; and on the means of
protecting buildings and shipping
against the destructive effects of
lightning. xvi, 226p. illus.
plate, front. London, 1843.
Mp195, 470; R; W1043.   FI, APS

784. ----. Protection of ships
     from lightning, according to
principles established by Sir W. S.
Harris...Compiled...by R. B. Forbes.
63, [1]p. Boston, 1848.
Not in MRW.   FI

785. ----. Rudimentary electri-
     city. [1st ed.]. 160p.
London, 1848.
R; W1143.   FI

786. ----. Rudimentary magne-
     tism. 3 pts. in 2v. illus.
London, 1850-52.
R; W1180.   FI

HAÜY, RENÉ JUST: see 238-42

HAZARD, EBENEZER: see 748

787. HENRICI, FRIEDRICH CHRISTOPH,
     b. 1795. Über die elektri-
cität der galvanischen Kette.
227p. Göttingen, 1840.
R.   FI

HENRY, JOSEPH: see 988, 1257-60

788. HÉRICART DE THURY, LOUIS
     ETIENNE FRANÇOIS, vicomte,
1776-1854. De l'influence des
arbres sur la foudre et ses effets
et considérations à ce sujet...
27p. Paris, 1838.
From Annales de l'agriculture fran-
çaise, 1838.
R; W2741.   APS

789. HICKMANN, JOHANN N
     Elektricität als prüfungs
und belebungsmittel im Scheintode.
84p. Vienna, 1841.
Not in MRW.   FI

790. HIGGINS, WILLIAM, 1763-1825.
     Experiments and observations
on the atomic theory, and electrical phenomena. 180p. Dublin,
1814.
W722.   UP

791. HIGGINS, WILLIAM MULLINGER.
     Alphabet of electricity, for
the use of beginners. viii, 116p.

illus. London, 1834.
R; W885.   UP

792. ----. The entertaining
     philosopher, a familiar ex-
planation of the most interesting
phenomena of natural and experimental philosophy... viii, 488p.
illus. London, 1844.
Not in MRW.   FI

793. HIGHTON, EDWARD.
     Specification of the patent
granted to Edward Highton for improvements in the electric telegraph, & in making telegraphic
communications. Sealed Feb. 7,
1850. 42p. plates. London, 1850.
Not in MRW.   FI

794. HIRSCHEL, ANASTASIUS.
     De electricitatis in corpus
animale vi et effectu. 39p.
Berlin, 1821.
Inaugural dissertation.
Not in MRW.   CP

795. HITT, J   D   .
     Hitt's electro-dynamics; or,
Electricity the universal cause of
motion in matter. 142p. plates.
Philadelphia, 1848.
Not in MRW.   FI

796. HOBART. MAGNETICAL and me-
     teorological observatory.
Observations...1841-48...under
superintendence of Edward Sabine.
3v. plates. London, 1850-53.
Presented by the British government.
Mp267; R.   APS

HODGKIN, THOMAS: see 695-96

797. HOPKINS, EVAN, d. 1867.
     On the connexion of geology
with terrestrial magnetism... vi,
[1], 129p. 24pl. London, 1844.
R; W1061.   FI, UP

798. HOPKINS, GEORGE F
     Observations on electricity,
looming, and sounds; together with
a theory of thunder showers, and

of west and north west winds. To
which are added, a letter from the
Hon. Thomas Jefferson, and remarks
by the Hon. Samuel L. Mitchill.
40p. New York, 1825.
Not in MRW.   HSP

799. HOWARD, LUKE, 1772-1864.
      Seven lectures on meteorol-
ogy... 2d ed. vi, 218p. illus.
London, 1843.
R: APS

see also: 696

HUMBOLDT, ALEXANDER, freiherr von:
      see 262-67, 660

IMHOF, MAXIMUS: see   269-70

800. IZARN, JOSEPH, 1766-1834.
      Manuale del galvanismo,
adatto alla fisica, alla chi-
mica e alla medicina. Tr. dal
francese. 206p. 6pl. Florence,
1805.
Mp617; R.   UP

801. ----. Manuel du galvanisme;
      ou, Description et usage des
divers appareils galvaniques employés jusqu'à ce jour tant pour
les recherches physiques que pour
les applications medicales. xxii,
304p. Paris, 1804.
C.2 (FI) dated 1805.
Autograph presentation copy from
author (UP).
R; W664.   FI, UP

JACKSON, CHARLES T.: see 811

802. JACOBI, MORITZ HERMANN, 1801-
      1874. Die Galvanoplastik,
oder das Verfahren cohärentes
Kupfer in Platten oder nach sonst
gegebenen Formen, unmittelbar aus
Kupferauflösungen, auf galvani-
schem Wege zu produciren...Nach
dem russischen Originale... [3],
viii, 63p. 1pl. St. Petersburg,
1840.
W982.   UP

803. ----. Galvanoplastik; or

The process of cohering copper into plates, or other given
forms, by means of galvanic action
on copper solutions...Tr. from the
German ed. by William Sturgeon.
vi, 39p. 1pl. Manchester, 1841.
W982a.   UP

804. JOHNSON, EDWARD JOHN, d. 1853.
      Report of magnetic experi-
ments tried on board an iron steam-
vessel... pp. 267-283. plate.
tables. London, 1836.
From Phil. Trans.
R; W2709.   APS

805. JOHNSON, MOSES.
      Brief and simple explanation
of the electro-magnetic telegraph;
its mode of operation. 24p. illus.
Cincinnati, 1847.
Not in MRW.   UP

806. JOHNSON, WALTER ROGERS,
      1794-1852. Observations on
the electrical characters of caout-
chouc, or gum elastic; with some
applications of which they are sus-
ceptible. 8p. n.p., [1830].
Presented by the author 1831.
Not in MRW.   APS

807. JOYCE, JEREMIAH, 1763-1816.
      Scientific dialogues...in
which the first principles of natural and experimental philosophy,
are fully explained...New ed. 3v.
24pl. Philadelphia, 1819.
(W746 & 746a).   APS

808. ----. Scientific dialogues:
      ...New ed. 576p. illus.,
front. Halifax, 1847.
(746 & 746a).   APS

808a. JUDEL, RENÉ FRANÇOIS.
      Considérations sur l'origine,
la cause et les effets de la fièvre,
sur l'électricité médicale et sur
le magnétisme animal. xii, 149p.
Paris & Versailles, 1808.
Corrections in manuscript.
Not in MRW.   APS

809. KAEMTZ, LUDWIG FRIEDRICH,

68

1801-1867. A complete course
of meteorology...tr., with notes
and additions by C. V. Walker.
xxii, 598p. 15pl. London, 1845.
R; W1076.   APS

810. KARSTEN, KARL JOHANN BERN-
     HARD, 1782-1853. Über con-
tact-Elektricität. Schreiben an
Herrn Alexander von Humboldt.
150p. plate. Berlin, 1836.
R; W908.   FI

811. [KENDALL, AMOS], 1789-1869.
     Morse's patent; full exposure
of Dr. Chas. T. Jackson's pretensions to the invention of the American electro-magnetic telegraph.
66p. n.p., [1850?].
W5013.   UP

812. ----. Morse's telegraph and
     the O'Reilly contract: the
violations of the contract exposed,
and the conduct of the patentees
vindicated. 35p. Louisville,
1848.
Not in MRW.   HSP

813. KLAPROTH, HEINRICH JULIUS
     von, 1783-1835. Lettre à
M. le Baron A. de Humboldt, sur
l'invention de la boussole. 138p.
3pl. Paris, 1834.
Mp622, etc; R; W886.   APS

814. KNOBLOCH, M   .
     Galvanismus in seiner tech-
nischen Anwendung seit dem Jahre
1840... 116p. Erlangen, 1842.
R; W1020.   FI

815. KOELLE, AUGUST, d. 1856.
     Ueber das Wesen und die
Erscheinung des Galvanismus. Oder,
Theorien des Galvanismus und der
geistigen Gährung nebst. Andeu-
tungen über den materiellen Zusam-
manhang der Naturreiche. viii,
303p. Stuttgart & Tübingen, 1825.
R.   APS

816. KUPFFER, ADOLPHE THEODOR,
     1799-1865. Receuil d'obser-
vations magnétiques faites à St.-

Pétersbourg et sur d'autres points
de l'empire de Russie. v, [2],
717p. 2pl. St. Petersburg, 1837.
Presented by the author 1839.
R.   APS

see also: 936

817. L   , G   A.
     A treatise on the motions of
the earth & magnet. 42p. Portsea,
England, 1849.
Not in MRW.   FI

818. LA BEAUME, MICHAEL.
     Du galvanisme appliqué à la
médecine...tr. de l'anglais, et
précédé de remarques, de considérations physiologiques, et d'observations pratiques sur le galvanisme,
par B.-R. Fabré-Palaprat. xxvi,
[9]-438p. Paris, 1828.
Mp385; R; W843.   CP

819. ----. On galvanism, with
     observations on its chymical
properties and medical efficacy in
chronic diseases with practical illustrations... xxvii, 271p. 3pl.
London, 1826.
Mp622; R.   CP

820. L'AMING, RICHARD.
     De l'application des axiomes
de la mécanique et du calcul géomé-
trique aux phénomènes de l'électricité. 28p. 1pl. Paris, 1839.
R; W961.   APS

821. LAMONT, JOHANN VON, 1805-1879.
     Ueber das magnetische observa-
torium der Königl. sternwarte bei
München. 56p. illus. Munich, 1841.
Not in MRW.   APS

822. LANGUTH, CARL JULIUS.
     De vi magnetis particulas
ferreas corpori vivo inflictas
extrahendi. 24p. Leipzig, 1833.
Inaugural dissertation.
Not in MRW.   CP

823. LAPOSTOLLE, ALEXANDRE, 1749-
     1831. Trattato sul modo di
preservare le abitazioni dal ful-

mine e le campagne dalla grandine.
Opera volgarizzata dal Francese
dal S. Antonio Bodei... [4], 189p.
2pl. Milan, 1821.
R.   APS

824. LARDNER, DIONYSIUS, 1793-
        1859. A manual of electri-
city, magnetism, and meteorology.
2v. illus., plates. London, 1841-
44.
V.2 edited & completed by Charles
V. Walker.
Forms v.9 & 10 of Lardner's Cabinet cyclopaedia.
Mp379, 390; R; W1062.   FI, UP,
                          LCP, APS

825. ----. Popular lectures on
        science and art; delivered
in the principle cities and towns
of the United States. 2v. illus.
2pl. front. New York, 1846.
Not in MRW.   APS

LA RIVE, ARTHUR AUGUSTE DE: see
        1025

826. LE BOUVIER-DESMORTIERS,
        URBAIN RENÉ THOMAS, 1739-
1827. Examen des principaux sys-
témes sur la nature du fluide électrique et sur son action dans les
corps organisés et vivants, 358,
[2]p. 2pl. port. Paris, 1813.
Mp591; R.   APS

827. LEIBL   .
        Praktische Beitrage zu den
Erfahrungen über den Magnet als
Heilmittel. 63p. Würzburg, 1832.
Inaugural dissertation.
Not in MRW.   CP

828. LEITCH, JOHN M   .
        Prospectus showing the cost
and comparative revenue of a line
of telegraph connecting New York
City with New Orleans. 8p. New
York, 1846.
Not in MRW.   UP

829. LEITHEAD, WILLIAM.
        Electricity, its nature,
operation & importance in the

phenomena of the universe. xiv,
399p. London, 1837.
Mp626; R; W924.   FI, UP

830. [LEREBOURS, NOËL MARIE PAYMAL],
        1807-1873. Traité de galvano-
plastie par J. L.... 122p. illus.
Paris, 1843.
R.   UP, APS

831. LETTER FROM the secretary of
        the treasury, [Geo. M. Bibb]
transmitting a letter from Professor Morse, relative to the magnetic telegraph. 18p. [Washing-
ton, 1844].
(28th Congress, 2d session, Doc.
No. 24.)
Not in MRW.   APS

832. LE YERRIER, M   .
        Electric telegraph for pri-
vate correspondence in France.
26p. n.p., 1850.
Not in MRW.   FI

LIEBIG, JUSTUS: see 867

833. LISICKI,   .
        De l'emploi du galvanisme
dans quelques cas de paralysie
partielle. 26p. Paris, 1837.
Inaugural dissertation.
Not in MRW.   CP

834. LLOYD, HUMPHREY, 1800-1881.
        An account of a method of de-
termining the total intensity of
the earth's magnetic force in absolute measure. 8p.
From the Proc. of the Royal Irish
academy, 1848.
Presented by the author.
R; W2883.   APS

835. ----. Account of the induc-
        tion inclinometer and of its
adjustments. 8p. London, 1842.
R.   APS

836. ----. Account of the magneti-
        cal observatory of Dublin and
of the instruments and methods of
observation employed there. 54p.
5pl. Dublin, 1842.

70

Presented by the author.
R; W2790.   APS

837.   ----.   On a new magnetical
        instrument for the measure-
ment of the inclination and its
changes. 16p. Dublin, 1842.
From Proc. of the Royal Irish academy.
Inscribed by the author.
R.   APS

838.   ----.   On the corrections re-
        quired in the measurement of
the magnetic declination. 11p.
tables. n.p., n.d.
From the Proc. of the Royal Irish
academy.
Presented by the author.
W2885.   APS

839.   ----.   On the determination
        of the intensity of the
earth's magnetic force in absolute
measure. 16p. Dublin, 1843.
From Trans. of the Royal Irish
academy.
Inscribed: from the author.
R.   APS

840.   ----.   On the induction of
        soft iron, as applied to the
determination of the changes of
the earth's magnetic force. 16p.
From the Proc. of the Royal Irish
academy, 1850.
R.   APS

841.   ----.   On the mutual action
        of permanent magnets, con-
sidered chiefly in reference to
their best relative position in an
observatory. 20p. plate. Dublin,
1840.
From the Trans. of the Royal Irish
academy.
R; W2769.   APS

842.   ----.   Results of observa-
        tions made at the magnetical
observatory of Dublin, during the
years 1840-43. First series --
Magnetic declination. 25p. 3pl.
tables. Dublin, 1849.
From Trans. of Royal Irish academy

v.XXII.
Inscribed by the author.
Not in MRW.   APS

843.   ----.   Supplement to a paper
        on the mutual action of per-
manent magnets. 10p. plate. Dublin, 1841.
From Trans. of the Royal Irish
academy.
R.   APS

see also: 941

844.   LOCKE, JOHN, 1792-1856.
        On the invention of the
electro-chronograph; a letter to
Nicholas Longworth. 75p. Cincinnati, 1850.
Not in MRW.   UP

845.   ----.   Report on the inven-
        tion & construction of his
electro-chronograph for the National
observatory... [6], xi, 67p. 17pl.
Cincinnati, 1850.
R.   APS

see also: 1261-63

846.   LOSTALOT-BACHOUÉ, JEAN PIERRE.
        Exposition d'un nouveau mode
de traitement des douleurs...2d éd.
Précédée d'une Nouvelle théorie de
la vie ou de l'action nerveuse...
199p. Paris, 1830.
Not in MRW.   CP

847.   LOUGHBOROUGH, JOHN.
        The Pacific telegraph and
railway, an examination of all the
projects for the construction of
these works... xx, [5]-80p. 2 maps.
St. Louis, 1849.
Not in MRW.   HSP

848.   LOVERING, JOSEPH, 1813-1892.
        An account of the magnetic
observations made at the observatory of Harvard University, Cambridge. pp.85-160. tables.
[Cambridge, 1845].
From Memoirs of the American Academy.
R.   APS

see also: 716

849. LUC, JEAN ANDRÉ DE, 1727-1817.
     Traité élémentaire sur le
fluide éléctrico-galvanique. 2v.
Paris & Milan, 1804.
R; W661.   APS

850. MACRERY, JOSEPH.
     On the principle of animation.
24p. Wilmington, 1802.
Inaugural dissertation.
Not in MRW.   CP

851. MAGAROTTO, ANTONIO.
     Franklini theoria de electri-
citatis principio in compendium re-
dacta et illustrata. 158p. 1pl.
Padua, 1805.
R.   UP, APS

852. MAGNETIC TELEGRAPH CO.
     Articles of association &
charter from the state of Maryland
together with the office regula-
tions & the minutes of the meetings of stockholders & board of
directors. 2v. New York & Baltimore, 1847-52.
Not in MRW.   FI

853. MAGRINI, LUIGI, 1802-1868.
     Nuovo motore elettro-magneti-
co invenzione di Luigi Magrini.
10p. 1pl. Padua, [1836].
From Annali delle scienze del
Regno Lombardo-Veneto, 1836.
R.   APS

854. ----. Sopra l'elettro-
     magnetismo e le recenti sco-
perte del Prof...Salvadore dal
Negro... 44p. 2pl. Padua, 1834.
R.   APS

855. ----. Telegrafo elettro-
     magnetico praticabile a
grandi distanze... 86 [2]p. 4pl.
Venice, 1838.
R; W940.   APS

856. MAISSAIT, MICHEL, 1770-1822.
     Mémoire sur quelques change-
ments faits à la boussole et au
rapporteur, suivi de la déscrip-
tion d'un nouvel instrument, nommé

grammomètre... 178p. 8pl. Paris,
1818.
Mp632; W749.   APS

857. MAJOCCHI, GIOVANNI ALESSANDRO,
     d. 1854. Istruzione teorica
e pratica sui parafulmini... 114,
[1]p. 1pl. Milan, 1826.
R.   APS

858. MAKERSTOUN OBSERVATORY.
     Observations in magnetism
and meteorology made...[1841-1846]
...Ed. by John Allan Broun. 5v.
illus., plates. Edinburgh, 1848-60.
Reprinted from Trans. of R. Soc. of
Edinburgh v.17-19.
R.   APS

see also: 647

MARKUS, M   .: see 665

MARUM, MARTIN VAN: see 354-359

859. MATTEUCCI, CARLO, 1811-1868.
     Influenza dell'elettricità
terrestre sui temporali. 14p.
Bologna, 1829.
R.   FI

860. ----. Lectures on the physi-
     cal phenomena of living beings
...tr. under the superintendence of
Jonathan Pereira. x, [15]-388p.
illus. Philadelphia, 1848.
(R); (W1064a).   APS, LCP

861. ----. Lezioni di fisica...
     date nella...Università di
Pisa. 3v. plates. Pisa, 1841-42.
R.   APS

862. ----. Traité des phénomènes
     électro-physiologiques des
animaux. Suivi des études anatomiques
sur le système nerveux et sur l'or-
gane électrique de la torpille par
Paul Savi. xix, 348p. 6pl. Paris,
1844.
Mp634; R; W1064.   FI, CP

MAYER, FEDERICO: see 1017

863. MEADE, WILLIAM, d. 1833.
     Outlines of the origin and

72

progress of galvanism; with its application to medicine. In a letter
to a friend. [2d ed.] 74, [3]p.
2pl. Dublin, 1805.
Presented by the author, 1816.
Mp285; R.   APS , CP

864. METCALFE, SAMUEL LYTLER,
       1798-1856. A new theory of
terrestrial magnetism. (Read before the New-York Lyceum of Natural
History.) 158p. New York, 1833.
R; W878.   FI, CP, APS

865. MOIGNO, FRANÇOIS NAPOLÉON
       MARIE, 1804-1884. Traité de
télégraphie électrique renfermant
son histoire, sa théorie & la dé-
scription des appareils. xxiv,
420p. plates. Paris, 1849.
R; W1161.   FI

866. MORIN, PYRAME LOUIS, 1815-
       1864. Essai sur la nature
et les propriétés d'un fluide im-
pondérable, ou, Nouvelle théorie
de l'univers matériel... xiv,
253, [2]p. Puy & Paris, 1819.
Autograph of author.
Not in MRW.   APS

MORSE, SAMUEL FINLEY BREESE:
     see 831, 882

MUELLER, C   H   .:
     see 963

867. MÜLLER, JOHANN HEINRICH
       JAKOB, 1809-1875. Kurze
Darstellung der Galvanismus; nach
Turner, mit Benutzung der Original
abhandlung Faraday's...Mit einem
Vorwort von J. Liebig. vi, 101p.
illus. Darmstadt, 1836.
R; W911.   UP

868. ----. Principles of physics
       and meteorology. x, 573p.
illus., front. London, 1847.
R; W1119.   LCP

869. NATURAL PHILOSOPHY. (Library
       of useful knowledge). 3v.
illus. London, 1829-34.
Not in MRW.   LCP

NEGRO, SALVATORE DAL: see 390-92,
     854

870. NOAD, HENRY MINCHIN, 1815-
       1877. A course of eight
lectures; on electricity, galvanism, magnetism, and electro-
magnetism. x, 382p. illus.,
front. London, 1839.
Bookplate: Essex Institute.
R.   APS

871. ----. Lectures on elec-
       tricity, comprising gal-
vanism, magnetism, electro-
magnetism, magneto- and thermo-
electricity. New ed. iv, [2],
457p. illus., front. London,
1844.
Mp228; R; W1065.   UP, LCP, APS

872. ----. Lectures on electri-
       city, comprising galvanism,
magnetism...& electrophysiology.
3rd ed. 505p. illus. London,
1849.
Not in MRW.   FI

873. NOBILI, LEOPOLDO, 1784-1834.
       Memoria su l'andamento e gli
effetti delle correnti elettriche
dentro le masse conduttrici. 51p.
illus. Florence, 1835.
R.   APS

874. ----. Memorie ed osserva-
       zioni edite ed inedite...
colla descrizione ed analisi de'
suoi apparati ed istrumenti. 2v.
in one. plates. Florence, 1834.
R; W887.   APS

875. ----. Nuovi trattati sopra
       il calorico, l'elettricità
e il magnetismo. viii, 401p. 8pl.
Modena, 1822.
R; (W943).   APS

see also: 892

876. OERSTED, HANS CHRISTIAN,
       1777-1851. Et capitel af
den elektromagnetiste probeerkunst.
441-456p. [Copenhagen], 1828.
From Magazin for kunstnere og

haandvaerkere.
(R).   APS

877. [----]. Experimenta circa
      effectum conflictus electrici
in acum magneticam. 4p. 1pl.
[Copenhagen, 1820.]
R.   APS

878. ----. Recherches sur l'iden-
      tité des forces chimiques et
électriques...Tr. de l'allemand par
Marcel de Serres. [3], xx, 258,
[2]p. plate. Paris, 1813.
Inscription: Donné par l'auteur le
2 juin 1813.
Ms. notes.
Mp453; R.   APS

879. OHM, GEORG SIMON, 1787-1854.
      Die galvanische Kette, mathe-
matisch bearbeitet. iv, 245p. 1pl.
Berlin, 1827.
Mp642; R; W835.   FI, UP, APS

880. OLMSTED, DENISON, 1791-1859.
      A compendium of natural
philosophy: adapted to the use of
the general reader, and of schools
and academies. xvi, 336p. illus.,
plate. New Haven, 1833.
Not in MRW.   APS

881. ----. An introduction to
      natural philosophy: designed
as a text book, for the use of the
students in Yale College...Compiled
from various authorities. 2v.
illus., plates. New Haven, 1832-35.
V.1 is 2d ed. 1835.
Not in MRW.   APS

882. O'RIELLY, HENRY, 1806-1886, &
      others vs. [Morse, S.F.B] &
others. A brief outline of the
progress of electric discovery,
with reference to telegraphic in-
ventions: and also a sketch of the
origin & progress of the "Atlantic,
Lake & Mississippi telegraph range
...as stated in the documents prepared for court," &c. v.p., [New
York], 1849.
Not in MRW.   UP

883. ORIOLI, FRANCESCO.
      De' paragrandini metallici.
Discorso quarto...Letto alla So-
cietà agraria di Bologna il giorno 16
marzo dell'anno 1826. 118p. Bologna, [1826].
R.   APS

884. OSANN, GOTTFRIED WILHELM,
      1797-1866. Grundzüge der Lehre
von dem Magnetismus und der Elektricität. viii, 183p. illus.
Würzburg, 1847.
R; W1120.   APS

885. PALLAS, EMMANUEL, b. 1792.
      De l'influence de l'électri-
cité atmosphérique et terrestre
sur l'organisme, et de l'effet de
l'isolement électrique considéré
comme moyen curatif et préservatif
...xii, 355p. Paris, 1847.
R.   CP

886. PALMER, W   .
      Electrotype: being a brief
description of the art of working
in metal by voltaic electricity.
27p. illus. London, 1841.
W2781.   FI

PALMIERI, LUIGI: see 1022

887. PARIS, JOHN AYRTON, 1785-
      1856. The life of Sir Hum-
phry Davy. 2v. illus., plate,
port. London, 1831.
Mp347; R; W861.   APS

888. PARK, ANDREW H   .
      A popular explanation of the
electric telegraph, and of some
other wonderful applications of
electricity. 34p. [Boston], 1849.
Not in MRW.   FI, HSP

889. PARK, ROSWELL, 1807-1869.
      Outline of magnetism. 20p.
n.p., n.d.
Not in MRW.   UP

890. PARTINGTON, CHARLES FREDERICK,
      d. 1857. A manual of natural
and experimental philosophy, being

the substance of a series of lec-
tures... 2v. illus., plates.
London, 1828.
Autograph: Robert Maskell Patterson.
W845.   APS

891. DE PAULA CANDIDO, FRANCISCO,
       1804-1864. Sur l'électricité
animale. 35p. Paris, 1832.
Inaugural dissertation.
Not in MRW.   CP

892. PELLI-FABBRONI, GIUSEPPE.
       Tributo di riconoscenza alla
memoria del cavalier professore
Leopoldo Nobili. 7p. [Florence,
1836].
Not in MRW.   APS

PEREIRA, JONATHAN: see 860

893. PERSON, CHARLES CLÉOPHAS,
       1801-1884. Théorie du gal-
vanisme. 37p. Paris, 1831.
Inaugural dissertation.
Mp645; R.   CP

894. PESCHEL, KARL FRIEDRICH,
       1793-1852. Elements of
physics. Tr. from German, with
notes by E. West. 3v. illus.
London, 1845-46.
Mp645; R; W1082.   FI

895. PETER, ROBERT, 1805-1894.
       On the application of gal-
vanic electricity to medicine.
Read to the Lexington medical society, December 2d, 1836. 19p.
n.p., n.d.
Reprint from Transylvania J. of
medicine...v.9, 1836.
Not in MRW.   CP

896. PETETIN, JACQUES HENRI DÉSIRÉ,
       1744-1808. Électricité ani-
male, prouvée par la découverte des
phénomènes physiques et moraux de
la Catalepsie hystérique...et par
les bons effets de l'électricité
artificielle dans le traitement de
ces maladies. xvi, 121, 382p.
port. Paris, 1808.
Includes "Notice historique sur la
vie et les ouvrages..." of author,

121p.
R.   APS

897. ----. Nouveau mécanisme de
       l'électricité, fondé sur les
lois de l'équilibre & du mouvement,
démontré par des expériences qui
renversent le système de l'électricité positive & négative; & qui
établissent ses rapports avec le
mécanisme caché de l'aimant dont
il explique les principaux phénomènes: et l'heureuse influence
du fluide électrique dans le
traitement des maladies nerveuses.
iv, xxviii, 300p. 10pl. Lyon,
1802.
R; W640.   APS

898. PEYTAVIN, JEAN-BAPTISTE.
       Nouvelle théorie de l'elec-
tricité, relativement aux corps
organisés, suivie d'un appendice
sur le somnambulisme magnétique.
iv, 98, [2]p. Nantes & Paris,
1826.
Not in MRW.   APS

899. PFAFF, CHRISTIAN HEINRICH,
       1773-1852. Der Electro-
Magnetismus, eine historish-
kritische Darstellung der bisheri-
gen Entdeckungen auf dem Gebiete
desselben, nebst eigenthümlichen
Versuchen. xiv, 288p. 8pl.
Hamburg, 1824.
M; R; W812.   APS

900. PIETTE, JACQUES EDME.
       Exposer les différentes mé-
thodes d'électrisation de l'homme
et leurs effets... 59p. Paris,
1843.
Inaugural dissertation.
Not in MRW.   CP

901. PINNOCK, WILLIAM, 1782-1843.
       Catechism of electricity,
being a short introduction to that
science; written in easy & familiar
language... 3rd ed. 72p. illus.,
port. London, [1822].
Not in MRW.   HSP, APS

see also: 921

902. POHL, GEORG FRIEDRICH, 1788-
1849. Der Elektromagnetismus und die Bewegung der Himmels-
körper in ihrer gegenseitigen
Beziehung. 95p. illus. Breslau,
1846.
R.   FI

903. ----. Der Process der galvanischen Kette. xxiv, 430p.
illus. Leipzig, 1826.
R; W825.   CP

POISSON, SIMÉON DENIS: see 738

904. POLLOCK, THOMAS.
Attempt to explain the phenomena of heat, electricity, galvanism, gravitation, and light, on
the assumption of one cause, or universal principle. xi, 158p. illus.
London, 1832.
W870.   UP, LCP

905. POMME, PIERRE, 1735-1812.
Réfutation de la doctrine
médicale de Docteur Brown...suivie
d'une notice sur l'électricité, le
galvanisme et le magnétisme, sous
le rapport des maladies nerveuses.
3rd ed. 144p. Arles, 1808.
Not in MRW.   CP

906. POPE, WILLIAM, inventor.
The triumphal chariot of
friction; or, A familiar elucidation of the origin of magnetic attraction, &c, &c. vii, 108p. 10pl.
London, 1829.
R; W851.   APS

907. PORTWINE, EDWARD.
Steam engine...Atmospheric
railways, electric printing telegraph, & screw-propeller. 2d ed.
144p. illus. London, 1847.
W1128.   FI

POUILLET, CLAUDE SERVAIS MATTHIAS:
see 696

908. PRAGUE. KAISERLICH königliche Sternwarte. Magnetische
und meteorologische Beobachtungen.
1840-1920. v.[2-21], 25-81.

Prague, 1842-1923.
Not in MRW.   APS

909. "PROGRESS, PETER" (pseud.)
The electric telegraph, comprising a brief history of former
modes of telegraphic communication,
an account of the electric clock,
Brett and Little's electro-tele-
graphic converser, Bain's printing
telegraph, etc. 84p. illus.,
front. London, 1847.
W1123.   LCP

910. DE PUISAYE, CHARLES.
De l'électricité considérée
comme moyen thérapeutique. 62p.
Paris, 1844.
Inaugural dissertation.
Not in MRW.   CP

911. QUETELET, LAMBERT ADOLPHE
JACQUES, 1796-1874. Note sur
le magnétisme terrestre suivie des
resultats des observations météoro-
logiques horaires faites à Bruxelles,
Louvain, Alost, et Londre. 11p.
n.p., n.d.
From Bull. de l'Acad. royale de
Brux. v.5, 1838.
Not in MRW.   APS

912. ----. Recherches sur les de-
grés successifs de force
magnétique qu'une aiguille d'acier
reçoit pendant les frictions multiples
qui servent à l'aimanter. 34p.
n.p., n.d.
From Annales de chimie et de physique, 1833.
Presented by author, 1834.
R.   APS

913. ----. Résumé des observations
sur la météorologie, sur le
magnétisme,...etc., faites à l'Observatoire royale de Bruxelles en
1840... 78p. Brussels, 1841.
From Mém. de l'Acad. royale de
Brux., v.14.
Not in MRW.   APS

914. ----. Sur l'emploi de la
boussole dans les mines.
34p. Brussels, 1843.

Presented by the author 1844.
From Annales des travaux publiques
de Belgiques.
R.   APS

915. READE, JOSEPH.
      A letter to Dr. Faraday re-
specting a hasty opinion, given...
on Dr. Joseph Reade's paper shewing
that radiant heat was converted into electricity by reflection. 12p.
illus. London, 1846.
Not in MRW.   FI

916. REICHENBACH, KARL LUDWIG
      FRIEDRICH, freiherr von,
1788-1869. Abstract of researches
on magnetism & certain allied subjects, including a supposed new im-
ponderable. tr. & abridged by William Gregory. 112p. plates.
London, 1846.
W1104.   FI, CP

917. ----. Physico-physiological
      researches on the dynamics of
magnetism, electricity, heat, light,
crystallization, and chemism, in
their relations to vital force...
from the German 2d ed...by John
Ashburner. 2v. illus. London,
1850-51.
(R).   LCP

918. ----. Researches on magne-
      tism, electricity, heat,
light, crystallization, & chemical
attraction, in their relations to
the vital force...Tr. & ed. by
William Gregory...Pts. I & II, including the 2d ed. of the first
part... xlv, 463p. 3pl. London,
[1850].
Mp140, W1188.   UP, FI, LCP, CP

919. RIADORE, JOHN EVANS, d. 1861.
      On the remedial influence of
oxygen or vital air, nitrous oxyde,
and other gases, electricity and
galvanism, in restoring the healthy
functions of the principle organs
of the body, and the nerves sup-
plying the respiratory, digestive,
and muscular systems. viii, viii,

177p. London, 1845.
Mp386, R.   CP

920. RITTER, JOHANN WILHELM, 1776-
      1810. Das electrische Sys-
tem der Körper: ein Versuch. 412p.
21 tables. Leipzig, 1805.
R; W673.   APS

921. ROBERTS, GEORGE, d. 1860?
      A catechism of electricity,
being a short introduction to that
science, written in easy and familiar language. Intended for the use
of young people. 68p. illus.,
front. London, [1820].
Not in MRW.   HSP

see also: 901

922. ROBERTS, MARTYN J   .
      The process of blasting by
galvanism, detailed in a letter addressed to the Highland and agri-
cultural society of Scotland.
36p. 3pl. London, 1840.
R; W988.   APS

923. ROBISON, JOHN, 1739-1805.
      A system of mechanical phi-
losophy. With notes by David
Brewster. 4v. plates. Edinburgh, 1822.
Mp311; R; W791.   APS

924. ROESSINGER, FRÉDERIC LOUIS.
      Résumé de l'ouvrage intitulé:
Fragment sur l'électricité univer-
selle, ou Attraction mutuelle.
41p. Geneva, 1840.
R.   CP

925. ROGERS, HENRY J   , 1811-
      1879. Telegraph dictionary
& seaman's signal book adapted to
signals by flags or other sema-
phores & arranged for secret correspondence through Morse's electro-
magnetic telegraph... 334p. plates.
Baltimore, 1845.
Not in MRW.   FI

926. ROGET, PETER MARK, 1779-1869.
      Treatises on electricity,

galvanism, magnetism, and electro-
magnetism...Published under the
superintendence of the Society for
the Diffusion of Useful Knowledge
(Library of Useful Knowledge).
v.p. illus. London, 1832.
Autograph: Franklin Bache.
Mp475; R; W871.   FI, CP, APS

927. ROSS, SIR JAMES CLARK, 1800-
       1862. On the position of the
north magnetic pole. pp.47-52. table.
London, 1834.
From Phil. Trans.
R; W2685.   APS

928. ----. A voyage of discovery
       and research in the southern
and antarctic regions, during the
years 1839-43. 2v. illus., plates.
London, 1847.
R.   APS

929. ROTH, JOHANN JOSEPH, 1786?-
       1866. De electricitatis in
organismum humanum effectu. vi,
21p. Munich, 1829.
Inaugural dissertation.
Not in MRW.   CP

930. ROYAL SOCIETY of London --
       Committee of physics &
meteorology. Proceedings. Nos.
1-3. Feb. 10-May 12, 1842. 35p.
n.p., n.d.
Not in MRW.   APS

931. ----. Report of a joint com-
       mittee of physics and mete-
orology referred to...for an opinion
on the propriety of recommending
the establishment of fixed magnetic observations in the Antarctic
Seas...together with the resolu-
tions adopted on that report, by
the Council of the Royal society.
7p. [London], n.d.
Not in MRW.   APS

932. ----. Report of the com-
       mittee of physics, including
meteorology, on the objects of
scientific inquiry in those
sciences. iv, 119, [1]p. 9 sample

record-keeping forms, 4pl. London,
1840.
Not in MRW.   FI, APS

933. ----. Report relative to
       the observations to be made
in the Antarctic expeditions & in
the magnetic observatories. [1],
119, [1]p. 4 charts. London, 1840.
Contains also blank forms for
books.
Presented by the Royal Society,
1840.
R; W994.   APS

934. ----. Revised instructions
       for the use of the magnetic
& meteorological observatories &
for the magnetic surveys. 47p.
tables. London, 1842.
R; W994a.   APS

935. ----. Supplemental instruc-
       tions for the use of the
magnetical observatories. 8p.
London, 1841.
Not in MRW.   APS

936. RUSSIA, OBSERVATOIRE physique
       central Nicolas. Annuaire
météorologique et magnétique du
corps des ingénieurs des mines...
[par] A.-T. Kupffer... 10v. plates.
St. Petersburg, 1837-49.
Continued as Annales...
(Mp623); R.   APS

937. RUTT, JOHN TOWILL, 1760-1841.
       Life and correspondence of
Joseph Priestley. 2v. London,
1831-32.
Not in MRW.   APS

938. SABINE, SIR EDWARD, 1788-1883.
       An account of experiments to
determine the figure of the earth,
by means of the pendulum vibrating
seconds in different latitudes; as
well as on various other subjects
of philosophical inquiry. xv, [1],
509, [2]p. 3 maps. London, 1825.
R.   FI, UP

939. ----. Contributions to terres-

78

trial magnetism. No. II.
[2], 11-35p. tables. London, 1841.
From Phil. Trans.
Inscribed by author.
R; W2784.   APS

940. ----. On the means adopted
        in the British colonial mag-
netic observatories, for determining the absolute values, secular
change, and annual variation of the
terrestrial magnetic force. pp.201-
219. London, 1850.
From Phil. trans.
Inscribed: from the author.
R.   APS

941. ----. Report on the magnetic
        isoclinal & isodynamic lines
in the British Islands. From obser-
vations by Professors Humphrey Lloyd,
and John Phillips; Robert Were Fox,
esq; Capt. James Clark Ross; and
Major Edward Sabine. 49-196p. 3pl.
London, 1839.
From 8th report of Br. Assn. for
Adv. of Sci.
Presented by the author.
R; W969.   APS

942. ----. Report on the phae-
        nomena of terrestrial mag-
netism: being an abstract of the
Magnetismus der Erde of Prof. Ch.
Hansteen. pp.61-90. 3pl. London,
1836.
From 7th Report of British assn.
for adv. of science.
Presented by Alexander Dallas Bache,
1837.
R.   APS

943. ----. Report on the varia-
        tions of the magnetic in-
tensity observed at different
points of the earth's surface.
85p. 5pl. London, 1838.
From 7th report of the British Assn.
for the Adv. of Science.
Inscribed by the author.
R; W945.   APS

944. ----. Terrestrial magnetism.
        2p. illus. plate (map).
Edinburgh, n.d. [1850].

(pp.71-72 of the Physical atlas of
natural phenomena, & plate 23).
Inscription: from Major-General
Sabine.
Not in MRW.   APS

see also: 745, 796, 945, 992

945. ST. HELENA, MAGNETICAL &
        meteorological observatory.
Observations, 1840-49...under superintendence of Edward Sabine. 2v.
plates. London, 1847-60.
Presented by the British government.
R.   FI(part), APS

946. SARLANDIÈRE, JEAN BAPTISTE,
        1787-1838. Mémoires sur
l'électropuncture considerée comme
moyen nouveau de traiter efficace-
ment la goutte, les rhumatismes et
les affections nerveuses,... iv,
150p. 1pl. (colored). Paris, 1825.
Mp385; R.   CP

SAVI, PAOLO: see 862

947. SCARSO, JOSEPH.
        Disputatio de magnete...tria
sumit publice defendenda Joseph
Scarso...adjutore Ludovico Menini
... 55p. Padua, 1813.
R.   APS

948. SCHELLEN, THOMAS JOSEPH HEIN-
        RICH, 1818-1884. Der elektro-
magnetische Telegraph in den ein-
zelnen Stadien seiner Entwicklung
und in seiner gegenwärtigen Ausbil-
dung und Anwendung, nebst einer
Einleitung über die optische und
akustische Telegraphie und einem
Anhange über die elektrischen Uhren.
xii, 368p. illus., plates. Braun-
schweig, 1850.
R.   FI, UP

949. SCHMIDT, CHRISTIAN HEINRICH.
        Handbuch der Galvanoplastik;
zunächt für Künstler und Gewerb-
treibende. 2d ed. 232p. plate.
Quedlinburg, 1847.
R.   FI

950. SCORESBY, WILLIAM, 1789-1857.
Essays on magnetism. v.p.
6pl. Edinburgh, 1832-33.
Contents: On the uniform perme-
ability of all known substances to
the magnetic influence; Exposition
of some of the laws & phenomena of
magnetic induction; Observations on
the deviation of the compass.
Inscription: from the Author.
R;(W2670).   APS

951. ----. Journal of a voyage to
the northern whale-fishery; including researches and discoveries
on the eastern coast of west Green-
land... xliii, 472p. plates, maps.
Edinburgh, 1823.
Presented by the author.
R.   APS

952. ----. Magnetical investiga-
tions. 4v. plates. London,
1839-52.
Inscription: from the author, 1844
(APS).
R.   FI, UP, APS(1&2), LCP(1)

953. SELMI, FRANCESCO, 1817-1881.
Nouveau manuel complet de
dorure et d'argenture par la méthode électro-chimique et par simple
immersion...tr. de l'italien...et
augmenté...par E. de Valicourt.
xii, 173p. Paris, 1845.
R.   FI, UP

954. SEYFFER, OTTO ERNST JULIUS,
b. 1823, Geschichtliche
Darstellung des Galvanismus.
638p. Stuttgart, 1848.
R.   FI

955. SHAW, GEORGE.
A manual of electro-metal-
lurgy. 49p. illus. London, 1842.
R; W1029.   FI

956. ----. Manual...2d ed. 202p.
illus. London, 1844.
R; W1029a.   FI

957. SHECUT, JOHN LINNAEUS EDWARD
WHITRIDGE, 1770-1836. Medical and philosophical essays, containing...[4 essays, including] An
inquiry into the properties and
powers of the electric fluid, and,
its artificial application to medical uses. The whole of which are
designed as illustrative of the
domestic origin of the yellow fever
of Charleston... viii, 260p. 1pl.
Charleston, 1819.
Not in MRW.   CP

958. SHERWOOD, HENRY HALL.
Electro-galvanic symptoms,
and electro-magnetic remedies, in
chronic diseases of the class
hypertrophy, or chronic enlarge-
ments of the organs and limbs, including all the forms of scrofula,
with illustrative diagrams and
cases. 3rd ed. 88p. New York,
1837.
Not in MRW.   LCP

see also: 996, 1219

959. SILLIMAN, BENJAMIN, 1779-
1864. Electro-magnetism.
History of Davenport's invention
of the application of electro-
magnetism to machinery; with remarks on the same from the American
journal of science & arts, by Prof.
Silliman. Also, extracts from
other public journals, & information on electricity, galvanism,
electro-magnetism, &c. by Mrs.
Somerville. 94p. New York, 1837.
Not in MRW.   FI, LCP

see also: 750

960. SILLIMAN, BENJAMIN, 1816-
1885. First principles of
chemistry, for the use of colleges
and schools...Fifteenth thousand.
[2d ed.]. 480p. illus. Philadelphia,
1850.
Not in MRW.   APS

961. SIMPSON, SIR JAMES YOUNG,
1811-1870. Observations re-
garding the influence of galvanism
upon the action of the uterus during
labour. 16p. tables. Edinburgh,
1846.

Reprinted from Monthly journal of
medical science, v.6.
R.   CP

962. SINGER, GEORGE JOHN, 1786-
      1817. Élémens d'électricité
et de galvanisme...tr. de l'anglais,
et augmenté de notes, par M. Thillaye.
viii, 655p. plates. Paris, 1817.
Mp430; R; W725a.   UP

963. ----. Elemente der Elektrici-
      tät und Elektrochemie...aus
dem Englischen...mit Anmerkungen
...von C. H. Mueller. xxv, 502p.
4pl. Breslau, 1819.
Mp430.   UP

964. ----. Elements of electri-
      city and electro-chemistry.
xxvii, 480p. 4pl. London, 1814.
Mp430; R; W725.   FI, UP, CP, APS

965. SMEE, ALFRED, 1818-1878.
      Elements of electro-biology,
or The voltaic mechanism of man;
of electro-pathology, especially
of the nervous system; and of
electro-theraputics. xii, 164p.
illus., 2pl. London, 1849.
Inscription: Amer. Phil. Soc.
Philad. from the author.
R; W1165.   FI, CP, APS

966. ----. Elements of electro-
      metallurgy or the art of
working in metals by the galvanic
fluid. xxviii,163p. illus.
London, 1841.
Mp660; R; W1006.   FI

967. ----. Elements of electro-
      metallurgy. 2d ed. 338p.
illus. London, 1843.
R; W1006a.   FI, UP

968. ----. Instinct and reason
      deduced from electro-biology.
xxxiv, 320p. illus. 10pl. (part
col.). London, 1850.
R.   LCP, CP

969. ----. Nouveau manuel com-
      plet de galvanoplastie, ou,
Éléments d'électro-metallurgie...

suivi d'un traité de daguerréo-
typie...publié par E. de Valicourt
...Nouvelle éd. (Manuels-Roret).
x, 566p. illus., plates. Paris,
1845.
R.   LCP

970. ----. On the intimate ra-
      tionale of the voltaic
force. 2d ed. pp.307-28. London, 1842.
Reprinted for private distribution
from Elements of electro-metallurgy.
(R; W1006a).   APS

971. ----. Principles of the hu-
      man mind deduced from physi-
cal laws; being a sequel to Elements of electro-biology; together
with the lecture on the voltaic
mechanism of man, delivered at
the London Institution, April 11,
1849. xvi, 16p. illus. London,
1849.
"Lecture on electro-biology, or the
voltaic mechanism of man...Reprinted
from 'The Lancet,' April 21st, 1849":
16p.
W1165a.   FI

972. ----. Principles of the
      human mind, deduced from
physical laws; together with a lecture on electro-biology, or the
voltaic mechanism of man. 64p.
illus. New York, 1850.
(W1165a).   LCP, CP

973. SMITH, FRANCIS ORMOND JONATHAN,
      1806-1876. The secret corre-
sponding vocabulary; adapted for
use to Morse's electro-magnetic
telegraph: and also in conducting
written correspondence, transmitted
by the mails, or otherwise. [230]p.
Portland, Maine, 1845.
Not in MRW.   APS

974. SMITH, SAMUEL B   .
      The medical application of
electro-magnetism. 96p. illus.
New York, 1850.
Not in MRW.   FI

975. SNOW, ROBERT, 1806-1854.

Observations of the aurora borealis. From September 1834 to September 1839. 17p. London, 1842.
Presented by the author.
R; W2795.   APS

976. SOMERVILLE, (MRS.) MARY (FAIRFAX), 1780-1872. On the connexion of the physical sciences ... 458p. London, 1834.
W890.   LCP

see also: 959

SPALLANZANI, LAZZARO: see 495-96

977. SPENCER, THOMAS.
     An account of some experiments made for the purpose of ascertaining how far voltaic electricity may be usefully applied to the purpose of working in metal. 26p. illus. Liverpool, 1839.
R.   FI

978. ----. Instructions for the multiplication of works of art in metal, by voltaic electricity. With an introductory chapter on electro-chemical decompositions by feeble currents. viii, 62p. illus. Glasgow, 1840.
W990.   LCP

979. STEINBERG, KARL, 1812-1852.
     Die Dynamide Elektricitaet, Magnetismus, Licht, Wärme. Verwandtschaftslehre und Stöchiometrie. [2], 83p. Berlin, 1846.
R; W1108.   LCP

980. STEINHEIL, KARL AUGUST, 1801-1870. Ueber Telegraphie, insbesondere durch galvanische Kräfte... 30p. 2pl. Munich, 1838.
R; W947.   LCP

981. STEVENSON, WILLIAM FORD, d. 1852. Most important errors in chemistry, electricity, and magnetism, pointed out and refuted; and the phenomena of electricity and the polarity of the magnetic needle accounted for and explained. 2d ed. 68p. London, 1847.
R; W1133.   LCP

982. STURGEON, WILLIAM, 1783-1850.
     Course of 12 elementary lectures on galvanism. xi, 231p. illus. London, 1843.
R; W1054.   FI

983. ----. Description of an electro-magnetic engine for turning machinery. pp.75-78. plate. [London, 1836].
From Annals of electricity, 1836.
Not in MRW.   APS

984. ----. Experimental and theoretical researches in electricity. First memoir. pp.17-44. 1pl. n.p., 1837.
Presented by author 1839.
W925.   APS

985. ----. Experimental and theoretical researches...
Second memoir. [79]-96p. 1pl. London, 1839.
From Trans. of London electrical society.
Presented by author, 1839.
W925.   APS

986. ----. Familiar instructions in the theory and practice of the beautiful art of electro-gilding & silvering...also, instructions in the formation and theory of voltaic batteries. 2d ed. 57p. illus. London, 1843.
Not in MRW.   UP

987. ----. Lectures on electricity delivered in the Royal Victoria Gallery, Manchester, during the session of 1841-42. xi, 240p. illus. London, 1842.
Mp662; R; W1032.   FI, UP

988. ----. On the electric shock from a single pair of voltaic plates, by Professor Henry...repeated, and new experiments... pp.67-75. 1pl. [London, 1836].
From the Annals of electricity, 1836.
Not in MRW.   APS

989. ----. Scientific researches, experimental & theoretical, in electricity, magnetism, galvanism, electro-magnetism, & electro-chemis-

try.   viii, 563p. plates. Bury,
1850.
Mp662; R; W1190.   FI

see also: 1024

990. SUE, PIERRE, 1739-1816.
      Histoire du galvanisme; et
analyse des différens ouvrages
publiés sur cette découverte, depuis son origine jusqu'à ce jour.
4v. 1pl. Paris, 1802-1805.
Presented by W.P.C. Barton, 1817.
Mp247, 249, etc.; R; W630.   APS

THILLAYE: see 962

991. THOMSON, THOMAS, 1773-1852.
      Outline of the sciences of
heat & electricity. xiv, 583p.
illus. London, 1830.
Mp665; R; W991.   FI, CP

THOUVENAL, PIERRE: see 512-514

992. TORONTO, CANADA. Magnetical
      & meteorological observa-
tory. Observations, 1840-48...
under superintendence of Edward
Sabine. 3v. London, 1845-57.
Presented by the British government.
Mp267; R.   FI, APS

993. A TREATISE on the nature and
      effects of heat, light, elec-
tricity, and magnetism, as being
only different developments of one
element. 91p. Cambridge, [Mass.],
1827.
Not in MRW.   APS

994. U.S. - NAVAL Observatory.
      Astronomical, magnetic, &
meteorological observations, 1845-
52, 1861-92. 39v. Washington,
1846-99.
Inscription: with Lt. Maury's com-
pliments to the Am. Philos. Society.
Not in MRW.   APS

995. U.S. - TREASURY Dept.
      Telegraphs for the United
States. Letter from the secretary

of the treasury... 37p. Washington, 1837.
Not in MRW.   FI

996. U.S. 25th CONGRESS. 2nd ses-
      sion. Committee on naval af-
fairs. Report...[on] the memorial
of Henry Hall Sherwood...claiming...
new and important discoveries in
magnetism generally & more particularly in the magnetism of the
earth and representing that he is
the inventor of...the geometer,
whereby...it is practicable and
easy,...to determine, merely by
the dip of the needle, the variation of the needle, and the latitude and longitude of any place...
23p. [Washington, 1838].
Not in MRW.   APS

997. U.S. 28th CONGRESS. 2nd ses-
      sion. House. Report no.
187. Magnetic telegraph from Baltimore to New York. 7p. Washington,
[1845].
Not in MRW.   FI

998. VAIL, ALFRED, 1807-1859.
      American electro-magnetic
telegraph, with the reports of
Congress & a description of all
telegraphs known employing electricity or galvanism. 208p. illus.
Philadelphia, 1845.
R; W1137.   FI, HSP, APS

999. ----. Description of the
      American electro-magnetic
telegraph: now in operation between
the cities of Washington and Baltimore. 24p. illus. Washington,
1845.
W1087.   HSP

1000. ----. Description of the
       American electro-magnetic
telegraph... 24p. illus. Washington, 1847.
W1087.   FI, APS, UP, HSP

1001. VANUXEM, LARDNER, 1792-1848.
      An essay on the ultimate
principles of chemistry, natural

philosophy, and physiology, deduced from the distribution of matter into two classes or kinds, and
from other sources. Part I. 91p.
Philadelphia, 1827.
Not in MRW.   APS

VASSALLI-EANDI, ANTONIO MARIA:
    see 527-28

1002. VEAU DE LAUNAY, CLAUDE JEAN,
    1755-1826. Manuel de l'élec-
tricité, contenant les principes
elementaire, l'exposition des sys-
têmes, la description et l'usage
des différens appareils électrique,
un exposé des méthodes employées
dans l'électricité médicale...
suivi d'une table chronologique de
tous les ouvrages relatifs a l'electricité. iv, 80, 272p. plates.
Paris, 1809.
Autograph: Robert Maskell Patterson.
Mp590; R; W695.   FI

1003. VÈNE, A
    Essai sur une nouvelle théorie
de l'électricité, contenant une
réfutation du systeme des deux
fluides vitré et résineux... 118p.
1pl. Arras, [ca. 1819?].
Pp.49-50 missing, replaced by ms.
copies (APS).
R; W752.   LCP, APS

VOLTA, ALESSANDRO: see 534-38, 631

1004. WALKER, CHARLES VINCENT, 1812-
    1882. Electrotype manipula-
tion. Pt. 1 - 22nd ed. Pt. 2 -
12th ed. 60, 62p. illus. London,
1849.
(Mp669); (R); (W1007).   FI

see also: 809, 824, 1026

1005. WALL, ARTHUR.
    On Wall's improvements in the
manufacture of iron, copper, steel
and other metals, by the application of voltaic electricity. ii,
56p. London, 1846.
R; W5509.   UP

1006. WASHINGTON & NEW ORLEANS

telegraph co. Articles of
association together with the
minutes of the meetings of the
stockholders & board of directors.
v.p. Washington, 1848.
Not in MRW.   FI

1007. WATKINS, FRANCIS.
    Popular sketch of electro-
magnetism or electro-dynamics...
and outlines of the parent sciences, electricity and magnetism.
iv, 83p. 3pl. London, 1828.
Presented by Horace C. Richards,
April 1942.
Mp484; R; W847.   FI, APS

1008. WEBER, GEORGES P   F
    Dynamologie organique et
thérapeutique, ou, Traité de l'influence physiologique et patho-
génique des fluides impondérables
... 47p. Paris, 1847.
Not in MRW.   CP

WEBER, JOSEPH: see 552-56, 1027

WEBER, WILHELM EDWARD: see 736

1009. WETZLER, JOHANN EVANGELIST,
    1774-1850. Beobachtungen
über den Nutzen und Gebrauch des
Keil'schen Magnet-electrischen
Rotations-Apparates in Krankheiten,
besonders in chronisch-nervösen,
rheumatischen und gichtischen...
[2], 182, [2]p. Leipzig, 1842.
R; W1034.   CP

WHITNEY, S   ALBERT: see 690

1010. WIGHTMAN, JOSEPH MILNER,
    1812-1885. Companion to
electricity: comprising a brief
history of the science; with a
description of the various kinds
of electrical machines & apparatus;
including a great variety of recent & interesting experiments.
63p. illus. Boston, 1843.
Not in MRW.   FI

WILKINSON, CHARLES HENRY:
    see 563-65

84

1011. ZAMBONI, GIUSEPPE, 1776-1846.
Della pila elettrica a secco.
Dissertazione. 55p. 3pl. Verona,
1812.
Mp420; R; W714.   LCP

1012. ZANTEDESCHI, FRANCESCO, 1797-
1873. Dei fenomeni elettrici
della machina di Armstrong e delle
cause loro assegnate dai fisici.
pp.567-590. [Venice, 1848?].
No t.p.
From Raccolta fisico-chimico italiana, 1848.
R.   APS

1013. ----. Del trasporto della
materia pesante nelle due
opposte correnti dell'apparato
voltiano della loro natura e del
moto vorticoso o a spirale dell'
arco luminoso. 11p. Vicenza, 1844.
From Annali delle scienze del Regno Lombardo-Veneto.
Presented by the author.
R.   APS

1014. ----. Della elettrotipia.
52p. 5pl. Venice, 1841.
R.   APS

1015. ----. Descrizione di una
macchina a disco per la
doppia elettricita e delle esperienze eseguite con essa com-
parativamente a quelle dell'elet-
tromotore voltiano. 13p. Venice,
1845.
From Mem. dell'I.R. Ist. Veneto di
scienze lettere, ed arti.
Presented by the author.
R.   APS

1016. ----. Elenco delle prin-
cipali opere scientifiche.
12p. Venice, 1849.
Presented by the author.
Not in MRW.   APS

1017. ----. Esperienze intorno
alle alterazioni della virtù
magnetica per l'azione del calorico
e di qualche altro fenomeno rela-
tivo; memoria dei...Francesco ab.
Zantedeschi...e Federico Mayer...

42p. 2pl. Verona, 1831.
Not in MRW.   APS

1018. ----. Lettera 1. Sul magneto-
telluro-elettrico in Italia
diretta al celebre elettricista
Carlo Walker di Londra. pp.453-458.
No t.p.
From Raccolta fisico-chemica in
italiana, 1846.
Not in MRW.   APS

1019. ----. Memoria sugli effetti
fisici, chimici e fisiologi-
ci prodotti dalle alternative delle
correnti d'induzione della macchina
elettro-magnetica di Callan. 8p.
n.p., [1845].
From Ann. delle scienze del Regno
Lombardo-Veneto.
Presented by the author.
R.   APS

1020. ----. Memoria sul termo-
elettricismo dinamico nei
circuiti formati di un solo metal-
lo. 15p. 1pl. Vicenza, 1844.
From Ann. delle scienze del Regno
Lombardo-Veneto.
Presented by the author.
R.   APS

1021. ----. Memoria sulle leggi
fondamentali che governano
l'elettro-magnetismo. 21p.
Verona, 1839.
R.   APS

1022. ----. Osservazioni alla
Descrizione della batteria
magneto-elettro-tellurica ed alla
continuazione delle ricerche intorno ai fenomeni d'induzione del
magnetismo terrestre de Luigi
Palmieri. [9]p. n.p., [1845].
From Rendiconto della R. Accad.
delle scienze di Napoli.
Presented by the author.
R.   APS

1023. ----. Trattato del magnetismo
e della elettricità. 2v. in
1. 7pl. Venice, 1844-45.
Inscription:...Omaggio dell'autore.
Mp423; R; W1090bis.   APS

Periodicals

1024. ANNALS OF electricity, magne-
       tism & chemistry, & guardian
of experimental science, conducted
by William Sturgeon. London. 1-10,
1836-43//
Mp662; R; W5860.   UP

1025. ARCHIVES DE l'électricité...
       Geneva. 1-5, 1841-45//
Mp491; R.   FI, APS

1026. ELECTRICAL MAGAZINE. London,
       1-2, 1843-46//
R.   FI, APS

1027. DER GALVANISMUS; eine Zeit-
       schrift von Professor Weber.
Landshut. v.1-4, 1802-3//
Mp285; R.   CP, APS

1028. LONDON ELECTRICAL society.
       Proceedings, 1841-3//
(Mp468); R; W5917.   FI, APS

1029. ----. Transactions, and
       proceedings, 1837-40//
R; W5918.   APS

ANIMAL MAGNETISM

Before 1801

1030. ABRÉGÉ DE l'histoire des
      magnétiseurs de Lyon, par
un nouveau converti. 8p. n.p.,
n.d. [17--].
      APS

1031. ABSONUS, VALENTINE.
      Animal magnetism. A ballad,
with explanatory notes and observations: containing several curious
anecdotes of animal magnetisers,
ancient as well as modern. 44p.
London, 1791.
      LCP

1032. L'ANTI-MAGNETISME martin-
      iste, ou Barbériniste; ob-
servations trouvées manuscrites
sur la marge d'une brochure intitulée: Réflexions impartiales
sur le magnétisme animal, faites
après la publication du Rapport
des Commissaires, &c. 43p.
[Lyons, 1784].
      APS

BAILLY, JEAN SYLVAIN: see 1059-
      61, 1067, 1070

BALSAMO, GIUSEPPE: see 1085

1033. [BARBARIN, CHEVALIER DE].
      Système raisonné du magnet-
isme universel. D'après les principes de M. Mesmer, Ouvrage au-
quel on a joint l'explication des
procédés du magnétisme animal ac-
comodés aux cures des différentes
maladies, tant par M. Mesmer que
par M. le Chevalier de Barbarin et
par M. de Puiségur, relativement
au somnambulisme; ainsi qu'une
notice de la constitution des

Sociétés dites de l'Harmonie...
Par la Société de l'Harmonie
d'Ostende. v, [3], 133p.
Ostende, 1786.
      APS

1034. BARBEGUIÈRE, JEAN BAPTISTE,
      1723-after 1797. La ma-
çonnerie mesmérienne, ou Les
leçons prononcées par Fr. Mocet,
Riala, Themola, Seca, & Célaphon,
de l'ordre des F. de l'harmonie,
en loge mesmérienne de Bordeaux,
l'an des influences 5784, & du
mesmérisme le 1er. Par Mr. J. B.
B---. 83p. Amsterdam, 1784.
      APS

1035. BELL, JOHN, Reverend.
      Animal electricity, and mag-
netism, &c. demonstrated after the
laws of nature; with new ideas upon
matter and motion, in two parts.
36, 44p. n.p., [ca. 1790].
      UP

1036. ----. An essay on somnam-
      bulism, or sleep walking,
produced by animal electricity and
magnetism, as well as by sympathy,
&c. 38p. Dublin, 1788.
      UP

1037. BELL, JOHN, 1763-1820.
      The general and particular
principles of animal electricity
and magnetism, &c. in which are
found Dr. Bell's secrets and
practice, as delivered to his
pupils... 80p. [London], 1792.
      LCP

1038. BERGASSE, NICOLAS, 1750-1832.

Considérations sur le magnétisme animal, ou sur la théorie du
monde et des êtres organisés, d'
après les principes de M. Mesmer...
avec des pensées sur le mouvement,
par M. le Marquis de Chatellux [i.e.
Chastellux]... 149p. The Hague,
1784.
Inscription: "à la Société har-
monique de Strasbourg (APS).
Benjamin Franklin's copy (HSP).
     HSP, APS

1039. [----]. Lettre d'un méde-
     cin de la Faculté de Paris
à un médecin du College de Londres; ouvrage dans lequel on
prouve contre M. Mesmer, que le
magnétisme animal n'existe pas.
70p. The Hague, 1781.
     APS

1040. BONNEFOY, JEAN BAPTISTE,
     1756-1790. Analyse rai-
sonée des rapports des commissaires chargés par le roi de
l'examen du magnétisme animal.
89p. [Lyon], 1784.
     APS

1041. ----. Analyse raisonée...
     98p. Lyon, 1784.
Benjamin Franklin's copy.
     HSP

1042. BOURZEIS, JACQUES AMABLE DE.
     Observation très-importante
sur les effets du magnétisme animal. 26, [2]p. Paris, 1783.
Benjamin Franklin's copy.
     HSP

1043. [BOUVIER, MARIE ANDRÉ JOSEPH].
     Lettres sur le magnétisme
animal; où l'on discute l'ouvrage
de M. Thouret, intitulé: Magnétisme
animal, & le rapport de MM. les
commissaires... 103, [3]p.
[Brussels], 1784.
     APS

1044. [CABANIS, PIERRE JEAN GEORGES],
     1757-1808. Serment d'un
médecin, prononcé le jour de sa ré-
ception, en 1783, dans des écoles

situées en face d'une église, et
près d'un hôpital. 6p. [Paris?,
1783].
Poem - defense of Mesmer.
     APS

CAGLIOSTRO, ALEXANDRO, CONTE,
     see 1085

1045. [CAMBRY, JACQUES], 1749-
     1807. Traces du magnétisme.
48p. front. The Hague, 1784.
     APS

CAULLET DE VEAUMOREL: see 1086

CHASTELLUX, FRANÇOIS JEAN, Mar-
     quis de: see 1038

CHASTENET: see PUYSÉGUR

1046. CONFESSION D'UN médecin,
     académicien et commis-
saire d'un rapport sur le magnétisme animale avec les remon-
trances et avis de son directeur.
70p. Paris, ca. 1785.
     CP

1047. [DAMPIERRE, ANTOINE ESMONIN,
     marquis de], 1743-1824.
Réflexions impartiales sur le
magnétisme animal, faites après
la publication du Rapport des commissaires... 50p. Geneva & Paris,
1784.
     APS

1048. [----]. Réflexions impar-
     tiales... 48p. Geneva &
Paris, 1784.
Benjamin Franklin's copy.
     HSP

see also: 1032

DESLON: see ESLON

1049. DEVILLERS, CHARLES, 1724-
     1809. Le colosse aux pieds
d'Argille. iv, 176p. [Paris],
1784.
     APS

1050. DIEU, L'HOMME et la nature.

Tableau philosophique d'une
somnambule. lvi, [14], 136p.
London, 1788.
This work appeared for the first
time in 1786 under the title:
Extrait du journal d'une cure
magnétique...
UP

1051. [ESLON, CHARLES D'], d. 1786.
Le magnétisme dévoilé. Par
M. Deslong, premier éleve de Mesmer.
12p. Paris, n.d.
No t.p.
CP

1052. ----. Observations sur les
deux rapports de MM. les
commissaires nommés par sa majes-
tie pour l'examen du magnétisme
animal... 31p. Philadelphia &
Paris, 1784.
Benjamin Franklin's copy (HSP).
HSP, APS

1053. ----. Observations sur le
magnétisme animal. 151p.
London & Paris, 1780.
CP, APS

1054. [----]. Supplément aux
deux rapports de MM. les
commissaires... 77, [3]p.
Amsterdam & Paris, 1784.
Benjamin Franklin's copy.
HSP

1055. [ESPREMESNIL, JEAN JACQUES
DUVAL D'], 1746-1794. Re-
flexions préliminaires a l'occasion
de la piece intitulée Les docteurs
modernes jouée sur le Théâtre
Italien, le seize Novembre 1784.
3p. [Paris, 1784].
APS

1056. [----]. Suite des réflexions
préliminaires à l'occasion
des Docteurs modernes. 8p. [Paris,
1784].
No t.p.
APS

see also: 1113-14

1057. EXTRAIT DES registres de
la faculté de médecine de
Paris. Du premier Décembre 1784.
8p. [Paris, 1784].
No t.p.
APS

EXTRAIT DU JOURNAL D'UNE CURE
MAGNÉTIQUE: see 1050

1058. [FONTETTE-SOMMERY, comte de].
Lettre à Monsieur D'Eslon...
27p. Glasgow & Paris, 1784.
Benjamin Franklin's copy.
HSP

1059. FRANCE. COMMISSION chargée
de l'examen du magnétisme
animal. Exposé des expériences
qui ont été faites pour l'examen
du magnétisme animal...par M.
Bailly, en son nom & au nom de Mrs.
Franklin, Le Roy, de Bory & Lavoisier, le 4 Septembre 1784. 15p.
Paris, Imprimerie Royale, 1784.
Benjamin Franklin's copy (HSP).
LCP, HSP, APS

1060. ----. Exposé des expéri-
ences... 15p. Paris, Chez
Moutard, 1784.
Benjamin Franklin's copy.
HSP

1061. ----. Exposé des expériences... pp.92-108. Paris,
1790.
In Bailly, J. S., Discours et
mémoirs, v.2.
APS

1062. ----. Rapport de l'un des
commissaires chargés par le
roi, du magnétisme animal. 51p.
Paris, 1784.
Benjamin Franklin's copy.
HSP

1063. ----. Rapport de l'un des
commissaires... 72p. Paris,
1784.
Signé A. L. de Jussieu.
In Recueil des pièces les plus interessantes...(q.v.).

90

Benjamin Franklin's copy (HSP).
    HSP, APS

1064. ----. Rapport des commis-
    saires chargés par le Roi de
l'examen du magnétisme animal.
66p. Paris, Imprimerie royale,
1784.
Signé B. Franklin, Majault, Le Roy,
Sallin, Bailly, d'Arcet, de Bory,
Guillotin, Lavoisier.
APS copies 2 and 3 are quarto.
    APS, LCP

1065. ----. Rapport des commis-
    saires chargés par le roi...
80p. Paris, Chez Gastelier, 1784.
    HSP, APS

1066. ----. Rapport des commis-
    saires chargés par le roi...
80p. Paris, Chez Moutard, 1784.
In Recueil des pièces les plus interessantes...(q.v.).
Benjamin Franklin's copy (HSP).
    UP, HSP, APS

1067. ----. Rapport des commis-
    saires chargés par le roi...
pp.[1]-91. Paris, 1790.
In Bailly, J. S., Discours et
memoires, v.2.
    APS

1068. ----. Rapport des commis-
    saires de la Société royale
de médecine, nommés par le Roi pour
faire l'examen du magnétisme animal. 11, 39p. Paris, Imprimerie
royale, 1784.
Pp.1-11 are letter from Mesmer to
le comte de C---.
Rapport signe Poissonnier, Caille,
Mauduyt, Andry.
Benjamin Franklin's copy (HSP) --
lacks pp.1-11.
    APS, HSP

1069. ----. Rapport des commis-
    saires de la Société royale
de médecine... 47p. Paris, Chez
Moutard, 1784.
Signé Andry, Caille, Mauduyt, et
Poissonnier-Desperrieres.
Benjamin Franklin's copy (HSP).
    APS, HSP

1070. ----. Rapport secret sur le
    mesmérisme. Redigé par Bailly.
pp.[146]-155. Paris, 1784.
"Fait à Paris le 11 abût 1784.
Signé Franklin, Bory, Lavoisier,
Bailly, Majault, Sallin, d'Arcet,
Guillotin, Leroi."
Photostat from Le Conservateur,
t.1 [1800].
    APS

1071. ----. Report of Dr. Benja-
    min Franklin, and other com-
missioners, charged by the King of
France, with the examination of
the animal magnetism, as now prac-
tised at Paris. Tr. from the
French with an historical introduction. xx, 108p. London, 1785.
Benjamin Franklin's copy (HSP).
    LCP, APS, HSP

1072. ----. Animal magnetism.
    Report of Dr. Franklin and
other commissioners,...Tr. from
the French. With an historical
outline of the "science," an abstract of the report on magnetic
experiments, made by a committee
of the Royal academy of medicine,
in 1831; and remarks on Col. Stone's
pamphlet. 2d ed. [2], 58p.
Philadelphia, 1837.
Contains also "Report of a committee of the Royal society of
medicine, appointed to examine a
work, entitled 'Enquiries and
doubts respecting the animal magnetism' by M. Thouret...To which
are subjoined by the translator,
notes, chiefly extracted from .
Thouret's performance": pp.1-8.
    UP, CP, HSP, APS

FRANKLIN, BENJAMIN: see 170-84,
    1059-61, 1064-67, 1070-72

FRENCH, BENJAMIN FRANKLIN: see
    1211

GALART DE MONTJOIE: see MONTJOIE

1073. [GIRARDIN, Dr.]
    Observations adressées a Mrs.
les Commissaires chargées par le
roi...sur la maniere dont ils y ont

procédé, & sur leur rapport.
Par un médecin de province. 36p.
London & Paris, 1784.
    APS

1074. ----. Observations ad-
    dressées à Messieurs les Com-
missaires de la Société royale de
médecine, nommées par le Roi...
Sur la manière dont ils ont procédé, et sur le rapport qu'ils en
ont fait. Par un médecin de P---.
Pour servir de suite à celles qui
ont été addressées sur le même objet à MM. les Commissaires tirés de
la Faculté de Médecine & de l'Académie royale des sciences de Paris.
17p. [London & Paris, 1784].
No t.p.
    APS

1075. HARSU, JACQUES DE, 1730-1784.
    Recueil des effets salutaires
de l'aimant dans les maladies. 60,
276p. Geneva, 1782.
    CP

1076. HERVIER, CHARLES, le père.
    Lettre sur la découverte du
magnétisme animal, a M. Court de
Gebelin... viii, 48p. Peking &
Paris, 1784.
    APS

1076a. [----]. Théorie du mes-
    mérisme. vi, 148p. Paris,
1818.
    APS

JUSSIEU, ANTOINE LAURENT: see
    1062-63

1077. [LAUGIER, ESPRIT-MICHEL].
    Parallele entre le magnét-
isme animal, l'électricité et les
bains médicinaux par distillation,
&c. appliqués aux maladies re-
belles...Par M. L.---, docteur en
médecine de l'Université de
Montpellier... 12, 91p. Paris,
1785.
Benjamin Franklin's copy.
    HSP

1078. LETTRE D'UN éleve de M.

Mesmer, a M. Pressavin. 16p.
[Lyon, 1784].
No t.p.
    APS

1079. LITTA BIUMI RESTA, CARLO
    MATTEO. Riflessioni sul
magnetismo animale...ad oggetto di
illuminare i suoi cittadini aven-
dolo trovato salutare in molti
mali. 234p. Italy [Milan?], 1792.
    APS

1080. [LUTZEBOURG,   COMTE DE].
    Cures faites par M. Le Cte.
De L........ sindic de la Société
de Bienfaisance etablie à Strasbourg...avec des notes sur les
crises magnétique appellées im-
proprement somnambulisme. 92p.
n.p., 1786.
Ms. notes.
    APS

1081. [----]. Extrait du journal
    d'une cure magnétique. Tr.
de l'allemand. [17], 136p.
Rastadt, 1787.
    APS

1082. LYONS. L'ÉCOLE Vétérinaire.
    Procès verbal de l'expéri-
ence magnétique faite...le lundi
9 Aout 1784, en présence de Monsieur le Comte d'Oels. 2p. Lyons,
1784.
Benjamin Franklin's copy.
    HSP

1083. MAGNETISM. CONSIDERED by
    Paracelsus & his followers
as the soul of the world! The in-
forming spirit of the universe!
The proper vehicle of the universal medicine!... 33p. n.p., n.d.
No t.p.
    FI

1084. MARTIN, JOHN, 1741-1820.
    Animal magnetism examined:
in a letter to a country gentleman.
69p. London, 1790.
    LCP

1085. MÉMOIRE POUR le Comte de

LETTRE

SUR LA DÉCOUVERTE

DU

MAGNÉTISME ANIMAL,

A M. COURT DE GEBELIN,

Cenſeur Royal, de diverſes Académies,
Préſident-Honoraire Perpétuel du Muſée
de Paris ;

Par le P. HERVIER, *Docteur de Sorbonne,*
Bibliothécaire des grands Auguſtins, &c.

Ad id ſufficit Natura quod poſcit.
La Nature ſuffit à ce qu'elle demande.
SENEC.

A PEKIN,

Et ſe trouve A PARIS,

Chez COUTURIER, Imprimeur-Libraire, Quai des
Auguſtins, près l'Egliſe, au Coq.

M. DCC. LXXXIV.

Figure 3. *Lettre sur la Decouverte du Magnetisme Animal, . . .* (See No. 1076)

Cagliostro, accusé; contre
M. le procureur-général, accusa-
teur; en présence de M. le Cardinal
de Rohan, de la Comtesse de la
Motte, & autres co-accusés. 53p.
Paris, 1786.
APS

1086. MESMER, FRIEDRICH ANTON,
1734-1815. Aphorismes de M.
Mesmer, dictés à l'assemblée de ses
eleves, & dans lesquels on trouve
ses principes, sa théorie & les
moyens de magnétiser...le tout for-
mant un corps de doctrine developpé
en 344 paragraphes,...ouvrages mis
au jour par M. C. de V. xxiv,
172p. Paris, 1785.
APS

1087. ----. Lettres...a Messieurs
les auteurs du Journal de
Paris, et a M. Franklin. 14p.
[Paris?, 1784].
APS

1088. ----. Mémoire sur la découverte du magnétisme animal. vi, 85p. Geneva, 1779.
CP, APS

1089. ----. Mémoire sur la decouverte du magnétisme animal. 468p. [Paris?], 1784.
In Recueil des pieces les plus
interessantes...(q.v.).
FI, APS

1090. ----. Mémoire...sur ses découvertes. xii, 110p. Paris,
[1799].
CP

1091. ----. Mémoires et aphorismes.
228p. [Paris, 1815].
CP

1092. ----. Précis historique des
faits relatifs au magnétisme-
animal jusques en Avril 1781. Tr.
de l'allemand. [6], 229p. London,
1781.
Two copies presented by the author
to APS.
APS, CP

1093. [----]. [Requête de M.
Mesmer au roi, sur la comé-
die des docteurs modernes; et
autres pieces relatives.] 4p.
n.p., [18--].
Ms. title page.
APS

1094. [----]. Systéme raisonné
du magnétisme universel.
D'après les principes de M. Mesmer...Par la Société d'harmonie
d'Ostende. v, [4], 133p. [Paris],
1786.
UP

see also: 507, 1033, 1068, 1216

1095. MONTJOIE, CHRISTOPH FELIX
LOUIS VENTRE DE LA TOULOUBRE,
1746-1816. Lettre sur le magnétisme
animal, où l'on examine la con-
formité des opinions des peuples
anciens & modernes, des sçavans, &
notamment de M. Bailly avec celles
de M. Mesmer; & où l'on compare
ces memes opinions au Rapport des
commissaires... viii, 136p.
Philadelphia & Paris, 1784.
CP, APS

1096. [MOUILLESAUX,   ], 1739-
1811. Appel au public sur
le magnetisme animal, ou projet
d'un journal pour le seul avan-
tage du public, et dont il serait
le coopérateur. 100p. n.p., 1787.
APS

1097. NOUVELLE DÉCOUVERTE sur le
magnétisme animal; ou, Lettre addressée à un ami de province,
par un partisan zélé de la verité.
64p. n.p., n.d.
APS

1098. PARIS, FACULTÉ de médecine.
Extractum é commentariis
saluberrimae facultatis Pariensis.
2p. [Paris], 1784.
HSP

1099. ----. Extrait des registres
de la Faculté de médecine de
Paris. 2p. [Paris], 1784.
Benjamin Franklin's copy.
HSP

1100. [PAULET, JEAN JACQUES], 1740-
1826. L'antimagnétisme, ou
Origine, progrès, décadence, re-
nouvellement et réfutation du magnétisme animal. [2], 252p. front.
London [& Paris], 1784.
APS, UP

1101. [----]. Mesmer justifié.
46p. Constance & Paris,
1784.
APS

1102. [PEARSON, JOHN], 1758-1826.
A plain and rational account
of the nature and effects of animal
magnetism: in a series of letters.
With notes and an appendix by the
editor. 51p. London, 1790.
CP

1103. [PETIAU,   ABBÉ].
Lettre de M. l'Abbé P.--- de
l'Académie de la Rochelle, à M.---
de la même Academie. Sur le magnétisme animal. 7p. n.p., [1784].
No t.p.
HSP

POISSONNIER-DESPERRIERES, ANTOINE,
see 1068-69

1104. A PRACTICAL display of...
animal magnetism, in which is
explained different modes of treat-
ing diseases...which it is pecu-
liarly adapted to relieve... 16p.
London, 1790.
FI

1105. PRESSAVIN, JEAN BAPTISTE.
Lettres sur le magnétisme.
16p. [Lyon, 1784].
No t.p.
APS

1106. PUYSÉGUR, ARMAND MARIE
JACQUES DE CHASTENET, marquis de, 1751-1825. Appel aux
savans observateurs du dix-neuv-
ième siècle, de la décision portée
par leurs prédécesseurs contre le
magnétisme animal, et fin du traitement du jeune Hébert. 11, 127p.
Paris, 1813.

See also his Continuation du journal...(1812) & Les fous...(1812);
the 3 were later printed as a single volume.
CP

1107. [----]. Continuation du
journal du traitement magnétique du jeune Hébert, [pendent
le] mois de septembre. 109p.
Paris, 1812
CP

1108. [----]. Détail des cures
opérées a Buzancy, près
Soissons, par le magnétisme animal.
42p. Soissons, 1784.
Benjamin Franklin's copy.
HSP

1109. ----. Du magnétisme animal,
considéré dans ses rapports
avec diverses branches de la physique générale. 478p. Paris, 1807.
APS

1110. ----. Les fous, les insen-
sés, les maniaques et les
frénétiques ne seraint-ils que des
somnambules desordonnés? 91p.
Paris, 1812.
CP

1111. ----. Mémoires pour servir
à l'histoire et à l'établis-
sement du magnétisme animal. 411p.
London, 1786.
UP, APS

1112. ----. Mémoires pour servir à
l'histoire...3rd ed. xxiv,
473p. front. Paris, 1820.
CP

see also: 1033

1113. PUYSÉGUR, JACQUES MAXINE
PAUL DE CHASTENET, comte de,
1755-1820. Rapport des cures
opérées a Bayonne par le magnétisme animal, adressé a M. L'Abbé
de Poulouzat...avec des notes de M.
Duval d'Esprémenil... 52p. Bayonne & Paris, 1784.
APS

1114. ----. Rapport des cures
        opérées a Bayonne... viii,
72p. Bayonne, 1784.
Benjamin Franklin's copy.
        HSP

see also: 1108

1115. RAPPORT DE la Société
        royale de médecine sur
l'ouvrage intitulée Recherches et
doutes sur le magnétisme animal,
etc. 22p. Paris, 1784.
        CP

1116. RÉCIT DE l'avocat-général
        de ---, aux chambres assem-
blées du public, sur le magnétisme
animal. 39p. Philadelphia & Paris,
1785.
        APS

1117. RECUEIL DES pieces les plus
        interessantes sur le magné-
tisme animal. 4 sep. vols. in 1.
[Paris], 1784.
Contains: Mesmer, Mémoire sur la
découverte du magnétisme animal,
1784. Rapport de l'un des commissaires...1784.--Rapport des commissaires...1784.--Rapport des commissaires de la société...1784.
Ea. vol. has sep. t.p. & paging.
        APS

1118. RECUEIL D'OBSERVATIONS et de
        faits relatifs au magnétisme
animal, présenté à l'auteur de cette
découverte, & publié par la Société
de Guienne. [2], 168p. Philadelphia, Paris, & Bordeaux, 1785.
        APS

1119. [RETZ DE ROCHEFORT, NOËL],
        d. ca. 1810. Memoire pour
servir à l'histoire de la jonglerie,
dans lequel on démontre les phénomenes du mesmérisme. [2], 47p.
front. London & Paris, 1784.
        APS

RHUBARBINI DE PURGANDIS, pseud.:
        see SERVAN

1120. [SALAVILLE, JEAN BAPTISTE],

1755-1832. Le moraliste mesmérien,
ou Lettres philosophiques sur l'influence du magnétisme. 132p.
London & Paris, 1784.
        APS

1121. [SERVAN, ANTOINE JOSEPH
        MICHEL], 1737-1807. Doutes
d'un provincial, proposées à M.M.
les médecins-commissaires, chargés
par le roi... 126p. [Lyon &
Paris], 1784.
        APS

1122. [----]. Doutes d'un provin-
        cial... 134p. Lyon & Paris,
1784.
        LCP, HSP

1123. [----]. Questions du jeune
        Docteur Rhubarbini de Purgan-
dis, adressés a Messieurs les doc-
teurs-regens, de toutes les fa-
cultés de médecine de l'univers,
au sujet de M. Mesmer & du magnétisme animal. xii, 50p. Padua,
1784.
        APS

SOCIÉTÉ DE GUIENNE: see 1118

SOCIÉTÉ DE L'HARMONIE D'OSTENDE:
        see 1033, 1094

1124. SOUSSELIER DE LA TOUR,
        L'ami de la nature, ou Mani-
ere de traiter les maladies par le
prétendu magnétisme animal. xiii,
175p. Dijon, 1784.
        APS

1125. STEARNS, SAMUEL, 1747-1819.
        The mystery of animal mag-
netism revealed to the world, containing philosophical reflections
on the publication of a pamphlet
entitled, A true and genuine discovery of animal electricity and
magnetism: also, an exhibition of
the advantages and disadvantages
that may arise in consequence of
said publication... [2], 58p.
London, 1791.
        HSP

1126. [TARDY DE MONTRAVEL, A   A].
Essai sur la théorie du som-
nambulisme magnétique. Par Mr.
T.D.M. xxii, 74p. London, 1786.
    APS

1127. [----]. Journal du traitement magnétique de la de-
moiselle N. Lequel a servi de base
à l'Essai sur la théorie du somnambulisme magnétique. xxxii,
255p. London, 1786.
    CP, APS

1128. THOURET, MICHEL AUGUSTIN,
1749-1810. Extrait de la
correspondance de la Société royale
de médecine, relativement au magnétisme animal. 74p. Paris, 1785.
    APS

1129. ----. Recherches et doutes
sur le magnétisme animal.
xxxv, [1], 251p. Paris, 1784.
    LCP, APS

see also: 11, 1043, 1072, 1115

1130. [TISSART DE ROUVRES, JACQUES
LOUIS NOËL, marquis de].
Nouvelles cures opérées par le
magnétisme animal. 64p. [Paris,
1784].
No t.p.
Benjamin Franklin's copy.
    HSP

VENTRE DE LA TOULOUBRE: see
    MONTJOIE

1801-1850

1131. ANIMAL MAGNETISM.
pp.388-415. Philadelphia,
1837.
(In American quarterly review, v.22).
    APS

AZAIS, PIERRE HYACINTHE: see 1132

1132. BAPST, F   G   .
Explication et emploi du
magnétisme; par MM. Bapst et Azais.
63p. Paris, 1817.
    APS

1133. BARTH, GEORGE H   .
The mesmerist's manual of
phenomena and practice; with directions for applying mesmerism to the
cure of diseases, and the methods
of producing mesmeric phenomena.
Intended for domestic use and the
instruction of beginners. 3rd ed.
viii, 206p. London, 1852.
    APS

1134. BEECHER, WILLIAM H   .
A letter on animal magnetism.
7p. Philadelphia, 1844.
    CP

see also: 1190

1135. BELL, JOHN, 1796-1872.
Animal magnetism: past fic-
tions-present science. 16p. Philadelphia, 1837.
Reprint from Select medical library
and eclectic journal of medicine.
    CP

1136. BERNA, DIDIER-JULES.
Expériences et considérations
à l'appui du magnétisme animal.
40p. Paris, 1835.
Inaugural dissertation.
    CP

1137. BERTRAND, ALEXANDRE JACQUES
FRANÇOIS, 1795-1831. Du magnetisme animal en France, et des
jugements qu'en ont portés les sociétés savantes avec le texte de
divers rapports faits en 1784...et
une analyse des dernières séances de
l'Academie royale de médecine et du
rapport de M. Husson; suivi de con-
siderations sur l'apparation de
l'extase dans les traitements magnétiques. xxix, 539p. Paris, 1826.
    APS

1138. BRAID, JAMES, 1795?-1860.

Neurypnology; or, The rationale of nervous sleep, considered
in relation with animal magnetism.
Illustrated by numerous cases of
its successful application in the
relief and cure of disease. xxii,
265, [1]p. London, 1843.
CP, APS

1139. BURDIN, CHARLES, c. 1778-1856.
Histoire académique du magnétisme animal accompagnée de notes
et de remarques critiques sur toutes
les observations faites jusqu'à ce
jour; par C. Burdin jeune et Fred.
Dubois (d'Amiens). xlvii, 651p.
Paris, 1841.
UP, LCP, CP

1140. BURNETT, CHARLES MOUNTFORD,
1807-1866. The philosophy of
spirits in relation to matter:
shewing the real existence of two
very distinct kinds of entity...by
which the phenomena of light, heat,
electricity, motion, life, mind,
etc. are reconciled and explained.
xx, 312p. London, 1850.
LCP

1141. BURTON, FRANCES BARBARA.
Elective polarity, the universal agent. viii, 171p. London,
1845.
LCP

1142. BUSH, GEORGE, 1796-1859.
Mesmer & Swedenborg; or, The
relation of the developments of
mesmerism to the doctrines and dis-
closures of Swedenborg. 288p.
New York, 1847.
LCP

1143. CALDWELL, CHARLES, 1772-1853.
Facts in mesmerism, and
thoughts on its causes and uses.
xxx, 132p. Louisville, 1842.
CP

1144. [CHARDEL, CASIMIR MARIE MAR-
CELLIN PIERRE CÉLESTIN],
1777-1847. Mémoire sur le magnétisme animal, présenté à l'Académie
de Berlin, en 1818. ii, 49p.

Paris, 1818.
Ms. notes.
APS

1145. CHARPIGNON, JULES i.e. LOUIS
JOSEPH JULES, b. 1815. Phy-
siologie, médecine et métaphysique
du magnétisme. 366p. Orléans &
Paris, 1841.
CP

1146. CHRISTMAS, HENRY, 1811-1868.
The cradle of the twin giants,
science and history. 2v. London,
1849.
LCP, APS

1147. COLLYER, ROBERT H
Psychography, or The embodi-
ment of thought; with an analysis of
phreno-magnetism, "neurology," and
mental hallucination, including
rules to govern and produce the
magnetic state. 44p. illus. Philadelphia, 1843.
HSP

1148. COLQUHOUN, JOHN CAMPBELL,
1785-1854. Isis revelata;
an inquiry into the origin, progress & present state of animal magnetism. 3rd ed. 2v. Edinburgh,
1844.
UP

1149. ----. Report of the experiments on animal magnetism,
made by a committee of the medical
section of the French royal academy
of sciences...tr. and now for the
first time published; with an historical and explanatory introduction, and an appendix. xii, 252p.
Edinburgh, 1833.
CP

1150. ----. Seven lectures on
somnambulism, tr. from the
German of Dr. Arnold Wienholt; with
a preface, introduction, notes, and
an appendix. xxxv, 219p. Edinburgh, 1845.
CP

see also: 1188

1151. COMMUN, JOSEPH DU.
Three lectures on animal magnetism, as delivered in New York,
at the Hall of science, on the 26th
of July, 2d and 9th of August.
78p. New York, 1829.
Presented by Alexander Dallas Bache,
(APS).
HSP, APS

1152. COTTON, CHARLES.
Popular delusions applied to
mesmerism, &c. 24p. Lynn, 1843.
APS

1153. CRUMPE, (MISS) G   S   .
Letters on animal magnetism.
23, [1]p. Edinburgh, 1845.
UP

1154. DELEUZE, JOSEPH PHILLIPPE
FRANÇOIS, 1753-1835. His-
toire critique du magnétisme animal. 2 parts in 1v. Paris, 1813.
APS

1155. ----. Instruction pratique
sur le magnétisme animal,
suivie d'un lettre écrite à l'auteur par un médecin étranger. [2],
472p. Paris, 1825.
CP

1156. ----. Practical instruction
in animal magnetism. Tr. from
the Paris ed. by Thomas C. Hartshorn.
[Part I] With notes by the translator referring to cases in this
country. 106, 36p. Providence,
1837.
CP

1157. ----. Practical instructions
...Tr. by Thomas C. Hartshorn.
Revised ed. With an appendix of
notes by the translator, and let-
ters from eminent physicians and
others, descriptive of cases in the
United States. 408p. New York &
Philadelphia, 1843.
CP

1158. ----. Practical instruction
... 415p. New York, 1846.
FI

1159. ----. Practical instruction
...Tr. by T. C. Hartshorn.
4th ed., with notes, and a life,
by Dr. Foissac. xxiv, 240p. London, 1850.
LCP

1160. DODS, JOHN BOVEE, 1795-1872.
The philosophy of electrical
psychology: in a course of twelve
lectures. Stereotype ed. 252p.
New York, [1850].
APS

1161. ----. Six lectures on the
philosophy of mesmerism, delivered in the Marlboro Chapel,
Boston. Reported by a hearer.
Tenth thousand. 82p. New York,
1849.
CP

DUBOIS, FRED. see 1139

1162. DUNCAN, GEORGE W   .
A practical treatise upon
mental philosophy, or, "Mesmerism."
...Compiled from...various works...
15p. Philadelphia, 1849.
HSP

1163. DUPAU, JEAN AMÉDÉE, b. 1797.
Analyse raisonée de l'ouvrage
intitulé: Le magnétisme éclairé, ou
Introduction aux Archives du magnétisme animal; par le baron d'Hénin
de Cuvillers... [20]-63p. [Paris,
1821].
From Revue médicale historique et
philosophique.
CP

1164. ----. Lettres physiologiques
et morales sur le magnétisme
animal, contenant l'exposé critique
des expériences les plus récentes,
et une nouvelle théorie sur ses
causes, ses phénomènes et ses applications à la médecine; addressées à M. le professeur Alibert.
xii, [2], 248p. Paris, 1826.
CP

1165. DU POTET DE SENNEVOY, JULES,
baron, 1796-1881. An intro-

duction to the study of animal magnetism...with an appendix, containing reports of British practitioners
in favour of the science. xi, 388p.
London, 1838.
          LCP

1166. ----. Manuel de l'étudiant
      magnétiseur, ou Nouvelle in-
struction pratique sur le magnétisme, fondée sur 30 années d'observation; suivi de la 4e éd. des Expériences faites en 1820 à l'Hôtel-
Dieu de Paris. xii, 344p. Paris,
1846.
          CP

1167. [----]. Le propagateur du
      magnétisme animal, par un so-
ciété de médecins. 2v. in 1.
Paris, 1827-28.
          APS

1168. ELLIOTSON, JOHN, 1791-1868.
      Numerous cases of surgical
operations without pain in the mesmeric state; with remarks upon the
opposition of many members of the
Royal medical and chirurgical society and others to the reception
of the inestimable blessings of mesmerism. 56p. Philadelphia, 1843.
          CP

ESCHENMAYER, CARL ADOLPH VON:
      see 1239

1169. ESDAILE, JAMES, 1808-1859.
      Mesmerism in India, and its
practical application in surgery
and medicine. xxxi, 287p. London, 1846.
          LCP

1170. ----. Mesmerism in India...
      259p. Hartford, 1847.
          CP

1171. FACTS IN magnetism, mesmer-
      ism, somnambulism, fascina-
tion, hypnotism, sycodonamy, ether-
ology, pathetism, &c., explained
and illustrated. Published for the
author. 96p. Auburn, 1849.
          LCP

1172. FILLASSIER, ALFRED.
      Quelques faits et considéra-
tions pour servir à l'histoire du
magnétisme animal. 91p. Paris,
1832.
Inaugural dissertation.
          CP

FOISSAC, DR.: see 1159

1173. FOISSAC, PIERRE, 1801-1886.
      Rapports et discussions de
l'Académie royale de médecine sur
le magnétisme animal recueillis par
un sténographe, et publiés, avec
des notes explicatives... 561p.
Paris, 1833
          CP

1174. FORBES, SIR JOHN, 1787-1861.
      Illustrations of modern mes-
merism from personal investigation.
xii, 101p. London, 1845.
          LCP, CP

1175. ----, ed. Mesmerism true --
      mesmerism false: a critical
examination of the facts, claims,
and pretensions of animal magnetism...with an appendix, containing a report of two exhibitions by
Alexis. 76p. London, 1845.
Reprinted from British & foreign
medical review.
          CP

FORMAN, J   G   :
      see 1241

1176. [FOURNEL, JEAN FRANÇOIS],
      1745-1820. Remonstrances des
malades aux médecins de la Faculté
Paris. 103p. Amsterdam, 1785.
          APS

1177. FRIEDLANDER, MICHAEL, 1769-
      1824. Lettre au rédacteur
de la Gazette de santé. Note sur
l'état actuel du magnétisme animal
en Allemagne. 11p. [Paris, 1817].
Extrait de la Gazette de santé,
no. 1.
          CP

1178. GORGERET, PHILIBERT MARIE

THE

HISTORY AND PHILOSOPHY

OF

ANIMAL MAGNETISM,

WITH PRACTICAL INSTRUCTIONS FOR THE EXERCISE OF THIS POWER.

BEING A COMPLETE COMPEND OF ALL THE INFORMATION NOW EXISTING UPON THIS IMPORTANT SUBJECT.

BY A PRACTICAL MAGNETIZER.

BOSTON:

PUBLISHED BY J. N. BRADLEY & CO,

OFFICE OF THE DAILY MAIL, 16 STATE STREET.

Figure 4. *The History and Philosophy of Animal Magnetism, . . .*

(See No. 1185)

EUZÈBE. Note sur le magnétisme et sur l'homéopathie; ou,
Réponse à tout ce qui a été imprimé
dans les journaux de Nantes contre
le magnétisme et contre l'homéopathie. 116p. illus. Nantes,
[not before 1841].
  CP

1179. GRANDVOINET, J   A   .
      Esquisse d'une théorie des
phénomènes magnetiques. 32p.
Paris, 1843.
      APS

1179a. GRIMES, JAMES STANLEY, 1807-
      1903. Etherology, and the
phreno-philosophy of mesmerism and
magic eloquence: including a new
philosophy of sleep and conscious-
ness, with a review of the pretensions of phreno-magnetism, electro-
biology, &c....Revised & edited by
W. G. Le Duc. 251, 121p. illus.
Boston & Cambridge, 1850.
      UP, LCP

1180. HADDOCK, JOSEPH W   .
      Psychology; or The science of
the soul considered physiologically
and philosophically, with an appendix, containing notes of mesmeric
and psychical experience. 109p.
illus. New York, 1850.
Published in London under title
Somnolism and psycheism...
      HSP, CP

1181. HAMARD, CHARLES PIERRE
      GUILLAUME. Expériences sur
le magnétisme animal. 18p.
Paris, 1835.
      CP

HARTSHORN, THOMAS C.: see 1156-59

1182. [HAYS, ISAAC], 1796-1879.
      On animal magnetism. -- By
the editor. 8p. n.p., n.d.
From the American journal of the
medical sciences, 1837.
      APS

1183. HÉNIN DE CUVILLIERS, ÉTIENNE
      FÉLIX, baron d', 1755-1841.

Magnétisme animal retrouvé dans
l'antiquité; ou, Dissertation historique etymologique et mythologique
... 2d ed. 432p. Paris, 1821.
      UP

1184. ----. Le magnétisme éclairé,
      ou introduction aux archives
du magnétisme animal. 252p. Paris,
1820.
Inscription & autograph of author.
      CP

see also: 1163, 1238

HERVIER, CHARLES: see 1076-1076a

1185. THE HISTORY and philosophy of
      animal magnetism, with prac-
tical instructions for the exercise
of this power, being a complete compend of all the information now
existing upon this important subject. By a practical magnetizer.
31p. Boston, [1843].
      HSP, APS

HUSSON,   : see 1137, 1214

1186. INCHBALD, MRS. ELIZABETH
      (SIMPSON), 1753-1821. Ani-
mal magnetism: a farce, in three
acts. 36p. New York, Aug. 1809.
      LCP

1187. [----]. Animal magnetism.
      A farce in three acts. As
performed at the Chestnut-Street
Theatre. 31p. Philadelphia, 1828.
      UP

1188. INSTITUT DE France - Académie
      des sciences. Report of the
experiments on animal magnetism,
made by a committee of the medical
section of the French royal academy of sciences:...Tr...with an
historical & explanatory introduction & an appendix by J. C. Colquhoun. xii, 252p. Edinburgh, 1833.
      UP, APS

1189. JOHNSON, CHARLES P   .
      A treatise on animal magne-
tism. 96p. New York, 1844.
      CP

1190. JONES, HENRY.
Mesmerism examined and re-
pudiated, as "the sin of witch-
craft," etc., especially in its
mysteries of clairvoyance...With
an appendix of mesmeric phenomena,
exhibited by William H. Beecher.
3rd ed. 24p. New York, 1846.
    HSP

1191. JOZWIK, ALBERT.
Sur le magnétisme animal.
13p. Paris, 1834.
Inaugural dissertation.
    CP

1192. A KEY to the science of electrical psychology. All its
secrets explained, with full and
comprehensive instructions in the
mode of operation, and its application to disease, with some useful
and highly interesting experiments
...By a professor of the science.
16p. n.p., 1849.
    HSP

1193. KLINGER, J   A   .
De magnetismo animali...
69p. Würzburg, 1817.
Inaugural dissertation.
    CP

1194. KLUGE, CARL ALEXANDER
FERDINAND. Versuch einer
Darstellung des animalischen Magne-
tismus, als Heilmittel. xiv, 612,
[2]p. [Berlin, 1811].
No t.p.
    CP

1195. ----. Versuch...3rd ed. xii,
356p. Berlin, 1818.
    CP, APS

1196. LAFONTAINE, CH   .
L'art de magnétiser; ou, Le
magnétisme animal considéré sous le
point de vue théorique, pratique et
thérapeutique. vii, 364p. Paris,
1847.
    CP

1197. LANG, WILLIAM.
Mesmerism; its history, phe-

nomena, & practice: with reports
of cases developed in Scotland.
xii, 240p. Edinburgh, 1843.
    UP

LE DUC, W   G   .:
    see 1179a

1198. LEE, EDWIN, d. 1870.
Animal magnetism and the as-
sociated phenomena, somnambulism,
clairvoyance, etc. 55p. London,
1849.
    FI

1199. ----. Observations on the
principal medical institu-
tions and practice of France, Italy
and Germany; with notices of the
universities, and cases from hospital practice. To which is added,
An appendix, on animal magnetism
and homeopathy. 102p. Philadelphia, 1837.
    CP

1200. LÉGER, THEODORE.
Animal magnetism, or psycho-
dunamy. 402p. New York & Philadelphia, 1846.
    UP, LCP

1201. [LOUBERT, JEAN BAPTISTE,
abbé]. Le magnétisme et le
somnambulisme devant les corps sa-
vants, le cour de Rome et les thé-
ologiens, par M. l'abbé J.-B. L.
702p. Paris, 1844.
    CP

1202. MAITLAND, SAMUEL ROFFEY,
1792-1866. Illustrations and
enquiries relating to mesmerism.
Part I. vi, 82p. London, 1849.
    CP

MARCARD,   DR.: see 1226

1203. MARTINEAU, HARRIET, 1802-
1876. Letters on mesmerism.
27p. New York, 1845.
    HSP

1204. ----. Letters on mesmerism.
xii, 70p. London, 1845.
    LCP, CP

1205. MESMERISM, OR, The new school
        of arts, with cases in point.
101p. London, 1844.
        LCP

1206. [MIALLE, SIMON].
        Exposé par ordre alphabétique
des cures opérées en France par le
magnétisme animal, depuis Mesmer
jusqu'à nos jours (1774-1826),
ouvrage où l'on a réuni les attes-
tations de plus de 200 médecins,
tant magnétiseurs que témoins, ou
guéris par le magnétisme. Suivi
d'un catalogue complet des ouvrages
français qui ont étés publiés pour,
sur ou contre le magnétisme. Par
M.S. 2v. Paris, 1826.
        CP

1207. MONTEGRE, ANTOINE FRANÇOIS
        JENIN DE, 1779-1819. Du mag-
nétisme animal et de ses partisans;
ou, Recueil de pièces importantes
sur cet objet, précédé des observations récemment. 139p. Paris,
1812.
        APS

1208. NEWNHAM, WILLIAM, 1790-1865.
        Human magnetism: its claims
to dispassionate inquiry: Being an
attempt to show the utility of its
application for the relief of human
suffering. 396p. New York, 1845.
        LCP, CP

1209. ----. Human magnetism;...
        vi, 432p. London, 1845.
        UP, LCP, CP

1210. PASLEY, T   H   .
        The philosophy which shows
the physiology of mesmerism, and
explains the phenomenon of clairvoyance. viii, 104p. London,
1848.
        LCP

1211. THE PHILOSOPHY of animal mag-
        netism: together with the
system of manipulating adopted to
produce ecstacy and somnambulism --
the effects and the rationale. By

a gentleman of Philadelphia...
84p. Philadelphia, 1837.
Attributed variously, e.g. to Edgar
Allan Poe, & to Benjamin Franklin
French.
        APS, CP, HSP

1212. PIGEAIRE, J   .
        Puissance de l'électricité
animale, ou du magnétisme vital et
de ses rapports avec la physique,
la physiologie et la médecine.
316p. Paris, 1839.
        APS

1212a. POE, EDGAR ALLAN, 1809-1849.
        Mesmerism "In Articulo Mortis."
An astounding & horrifying narrative, shewing the extraordinary
power of mesmerism in arresting the
progress of death. 16p. London,
1846.
        UP

see also: 1211

1213. POYEN SAINT SAUVEUR, CHARLES.
        A letter to Col. Wm. L. Stone,
of New York, on the facts related
in his letter to Dr. Brigham, and
a plain refutation of Durant's exposition of animal magnetism, &c.
by Charles Poyen. With remakrs on
the manner in which the claims of
animal magnetism should be met and
discussed. By a member of the
Massachusetts bench. 72p. Boston,
1837.
        HSP

1214. ----. Report on the magne-
        tical experiments made by the
Commission of the Royal academy of
medicine of Paris, read in the
meetings of June 21 and 28, 1831,
by Mr. Husson...Tr. from the French
& preceded with an introduction...
172p. Boston, 1836.
        UP

1215. REESE, DAVID MEREDITH, 1800-
        1861. Humbugs of New-York:
being a remonstrance against popular delusions; whether in science,

philosophy, or religion. 2d ed.
xii, [13]-267, [1]p. New York &
Boston, 1838.
        APS

REICHENBACH, KARL LUDWIG FRIEDRICH,
        freiherr von: see 916-18

1216. RICARD, J   J   A   .
        Physiologie et hygiene du
magnétiseur; régime diététique du
magnétisé; Mémoires et aphorismes
de Mesmer, avec des notes. xii,
216, 228p. Paris, 1844.
        CP

1217. ----. Traite théorique et
        pratique du magnétisme ani-
mal, ou, Méthode facile pour ap-
prendre à magnétiser. xii, 556p.
Paris, 1841.
        UP, CP

1218. SANDBY, GEORGE, 1799-1880.
        Mesmerism and its opponents:
with a narrative of cases. x, 278p.
London, 1844.
        UP, LCP, APS

1219. SHERWOOD, HENRY HALL.
        Motive power of organic life,
and magnetic phenomena of terrestrial and planetary motions, with
the application of the ever-active
and all-pervading agency of magnetism to the nature, symptoms and
treatment of chronic diseases.
196p. plates. New York, 1841.
        CP

see also: 958, 996

1220. SIMON, CLAUDE GABRIEL.
        Mémoire sur le magnétisme
animal et sur son application au
traitemant des maladies mentales
... 19p. [Nantes, 1834].
        APS

1221. SMITH, GIBSON.
        Lectures on clairmativeness:
or, Human magnetism. With an appendix. 40p. New York, 1845.
        APS

1222. SMITH, HORATIO, 1779-1849.
        Love and mesmerism...by
Horace Smith. 3v. London, 1845.
        UP

1223. [SOUTH, THOMAS].
        Early magnetism in its higher
relations to humanity, as veiled in
the poets and the prophets. By ΘΥΟΣ
ΜΑΘΟΣ viii. 127p. illus., front.
London, 1846.
        LCP

1224. STONE, WILLIAM LEETE, 1792-
        1844. Letter to Dr. A. Brig-
ham on animal magnetism, being an
account of a remarkable interview
between the author & Miss Loraina
Brackett while in a state of somnambulism. 66p. New York, 1837.
        CP, HSP, APS

1225. ----. Letter to Dr. A.
        Brigham...2d ed. 75p. New
York, 1837.
        UP, CP, APS

see also: 1072

1226. STROMBECK, FREDERIC CHARLES,
        baron de, 1771-1848. His-
toire de la guérison d'une jeune
personne, par le magnétisme animal, produit par le nature elle-
même. Par un témoin oculaire de
ce phénomène extraordinaire. Tr.
de l'allemand...avec un préface
du Dr. Marcard... 200, [1]p.
Paris, 1814.
        CP, APS

SUTHERLAND, LA ROY: see 1240

SWEDENBORG, EMANUEL: see 1142

1227. TESTE, ALPHONSE, b. 1814.
        Le magnétisme animal ex-
pliqué, ou Leçons analytiques sur
la nature essentielle du magnétisme, sur ses effets, son histoire, ses applications, les diverses manières de la pratiquer,
etc. [3], 479p. Paris, 1845.
        CP

1228. ----. Manuel pratique de
    magnétisme animal. Exposi-
tien méthodique des procédés employés pour produire les phénomènes
magnétiques et leur application à
l'étude et au traitement des maladies. viii, 476p. Paris, 1840.
    CP

1229. ----. Manuel pratique...
    3rd ed. viii, 500p. Paris,
1846.
    CP

1230. ----. Practical manual of
    animal magnetism; containing
an exposition of the methods employed in producing the magnetic
phenomena; with its application to
the treatment and cure of diseases.
Tr. from the 2d ed. by D. Spillan.
xii, 406p. London, 1843.
    UP, CP

1231. ----. Practical manual...
    Tr. from the 2d ed. by D.
Spillan. 1st Amer. ed. xv, 321p.
Philadelphia, 1844.
    CP

1232. TONNA, (MRS.) CHARLOTTE
    ELIZABETH (BROWNE), 1790-
1846. Mesmerism, a letter to Miss
Martineau. 16p. Philadelphia,
1847.
    UP

1233. TOWNSHEND, CHAUNCEY HARE,
    1798-1868. Facts in mes-
merism, with reasons for a dispassionate inquiry into it. xii,
575p. London, 1840.
    LCP

1234. ----. Facts in mesmerism...
    388p. New York, 1841.
    UP, LCP

1235. ----. Facts in mesmerism...
    2d ed. xxxv, 390p. London,
1844.
    APS

1236. ----. Facts in mesmerism or

animal magnetism. With reasons for a dispassionate inquiry
into it...1st Amer. ed., with an
appendix, containing the report of
the Boston committee on animal magnetism. x, [1], 539p. Boston,
1841.
    CP

1237. WESERMANN, H   M   .
    Der Magnetismus und die all-
gemeine Weltsprache. viii, 271p.
Creveld & Cologne, 1822.

WIENHOLT, ARNOLD: see 1150

Periodicals

1238. ARCHIVES DU magnétisme ani-
    mal. Paris.
1-8, 1820-23//
    UP

1239. ARKIV FÜR den thierischen
    Magnetismus in Verbindung mit
mehreren Naturforschern. hrsg. von
C. A. von Eschenmayer... Altenburg.
1-12, 1817-23.
    UP

1240. THE MAGNET. by La Roy Suther-
    land. New York.
v.1, no.7-9; v.2, no.3, 8. Dec.
1842-Feb. 1844.
    HSP

1241. PHRENO-MAGNETIC vindicator,
    ed. & pub. by J. G. Forman.
Lexington.
v.1, no.1 October, 1842.
    LCP

1242. ZOIST: A journal of cerebral
    physiology & mesmerism.
London.
1-13, March 1843-Jan. 1856//
    UP

ARTICLES ON ELECTRICITY AND MAGNETISM
IN AMERICAN PHILOSOPHICAL SOCIETY TRANSACTIONS BEFORE 1850

1243. BACHE, ALEXANDER DALLAS, 1806-1867. Observations of the magnetic intensity at twenty-one stations in Europe. pp.75-99. tables. (n.s.v.7, 1841.)

1244. ----. Observations to determine the magnetic dip at Baltimore, Philadelphia, New York, West Point, Providence, Springfield, and Albany. By A. D. Bache ...and Edward H. Courtenay. pp.209-215. tables. (n.s.v.5, 1837.)

1245. ----. On the diurnal variation of the horizontal needle. pp.1-21. tables, 3pl. (n.s.v.5, 1837.)

1246. ----. On the relative horizontal intensities of terrestrial magnetism at several places in the United States, with the investigation of corrections for temperature, and comparisons of the methods of oscillation in full and in rarefied air. pp.427-457. tables. (n.s.v.5, 1837.)

1247. BRYANT, WILLIAM, 1730-1786. Account of an electrical eel, or the torpedo of Surinam. pp.166-169. (v.2, 1786.)

1248. FLAGG, HENRY COLLINS. Observations on the numb fish, or torporific eel. pp.170-173. (v.2, 1786.)

1249. FRANKLIN, BENJAMIN, 1706-1790. Letter to Mr. Nairne of London, proposing a slowly sensible hygrometer for certain purposes. Nov. 13, 1780. pp.51-56. (v.2, 1786.)
Some references to electricity.

1250. ----. Queries and conjectures relating to magnetism, and the theory of the earth, in a letter...to Mr. Bodoin [James Bowdoin]. pp.10-13. [v.3, 1793.)

see also: 170-84, 519, 1059-61, 1064-67, 1070-72, 1286

1251. GRAHAM, JAMES DUNCAN, 1799-1865. Observations of the magnetic dip, made at several positions, chiefly on the south-western and north-eastern frontiers of the United States, and of the magnetic declinations at two positions on the River Sabine; in 1840. pp.329-80. tables. (n.s.v.9, 1846.)

1252. HARE, ROBERT, 1781-1858. Account of a tornado, which, towards the end of August 1838, passed over the suburbs of the city of Providence...and afterwards over a part of the Village of Somerset. Also an extract of a letter on the same subject from Zachariah Allen... pp.297-301. (n.s.v.6, 1839.)

1253. ----. Description of an electrical machine, with a plate four feet in diameter, so constructed as to be above the operator: also of a battery discharger employed therewith: and some observations on the causes of the diversity in the length of the sparks erroneously distinguished by the terms positive and negative. pp.365-73. illus., 1pl. (n.s.v.5, 1837.)

1254. ----. Engraving and description of a rotatory multiplier, or one in which one or more needles are made to revolve by a galvanic current. pp.343-45. illus. (n.s. v.6, 1839.)

1255. ----. On the causes of the tornado, or water spout. pp.375-84. illus. (n.s.v.5, 1837.)

<u>see also</u>: 759-82

1256. HASSLER, FERDINAND RUDOLPH, 1770-1843. Description of the magnetic needles. pp.354-56. (n.s.v.2, 1825.)

1257. HENRY, JOSEPH, 1797-1878. Contributions to electricity and magnetism...No. I.--Description of a galvanic battery for producing electricity of different intensities. pp.217-22. 1pl. (n.s.v.5, 1837.)

1258. ----. Contributions...No.II. On the influence of a spiral conductor in increasing the intensity of electricity from a galvanic arrangement of a single pair, &c. pp.223-31. 1pl. (n.s.v.5, 1837.)

1259. ----. Contributions...No.III. --On electro-dynamic induction. pp.303-37. illus. (n.s.v.6, 1839.)

1260. ----. Contributions...No.IV. --On electro-dynamic induction. (Continued) pp.1-35. illus. (n.s.v.8, 1843.)

<u>see also</u>: 988

1261. LOCKE, JOHN, 1792-1856. Observations made in the years 1838, '39, '40, '41, '42, and '43, to determine the magnetical dip and the intensity of magnetical force, in several parts of the United States. pp.283-328. tables. 4pl. (ns.sv.9, 1846.)

1262. ----. Observations to deter-

mine the horizontal magnetic intensity and dip at Louisville, Kentucky, and at Cincinnati, Ohio. pp.261-64. tables. (n.s.v.7, 1841.)

1263. ----. On the magnetic dip at several places in the State of Ohio, and on the relative horizontal magnetic intensities of Cincinnati and London...In a letter to John Vaughan...May 7, 1838. pp.267-73. tables. (n.s.v.6, 1839.)

1264. LOOMIS, ELIAS, 1811-1889. Additional observations of the magnetic dip in the United States. pp.101-111. tables. (n.s. v.7, 1841.)

1265. ----. Observations of the magnetic dip in the United States. Fourth series. pp.284-304. tables. (n.s.v.8, 1843.)

1266. ----. Observations to determine the magnetic dip at various places in Ohio and Michigan...In a letter to Sears C. Walker, Esq., M.A.P.S. pp.1-6. table. (n.s.v.7, 1841.)

1267. ----. Observations to determine the magnetic intensity at several places in the United States, with some additional observations of the magnetic dip. pp.62-71. tables. (n.s.v.8, 1843.)

1268. MADISON, JAMES, Bp., 1749-1812. Experiments upon magnetism. Communicated in a letter to Thomas Jefferson, President of the Philosophical Society... pp.323-28. (v.4, 1799.)

1269. NICOLLET, JOSEPH NICOLAS, 1786-1843. Observations of the magnetic dip, made in the United States, in 1841. pp.315-326. tables. (n.s.v.8, 1843.)

1270. OLIVER, ANDREW, 1731-1799. A theory of lightening and thunder storms. pp.74-101. (v.2, 1786.)

1271. ----. Theory of water spouts.
pp.101-117. (v.2, 1786.)

1272. PAGE, JOHN, 1743?-1808.
Account of a meteor...[in a
letter to David Rittenhouse, with
the latter's reply]. pp.173-76.
(v.2, 1786.)

1273. PATTERSON, ROBERT, 1743-1824.
An easy and accurate method
of finding a true meridian line,
and thence the variation of the
compass. pp.251-59. tables. [v.2,
1786.)

1274. ----. An improvement on
metallic conductors or light-
ening-rods, in a letter to Dr.
David Rittenhouse,... pp.321-24.
(v.3, 1793.)

1275. RITTENHOUSE, DAVID, 1732-
1796. An account of the
effects of a stroke of lightning on
a house furnished with two conductors,-in a letter from Messrs.
David Rittenhouse and Francis Hopkinson; to Mr. R. Patterson.
pp.122-25. (v.3, 1793.)

1276. ----. Account of several
houses in Philadelphia,
struck with lightning, on June 7,
1789. By Mr. David Rittenhouse,
and Dr. John Jones. pp.119-22.
plate. (v.3, 1793.)

1277. ----. An account of some experiments on magnetism, in a
letter to John Page, esquire, at
Williamsburg. pp.178-81. (v.2,
1786.)

1278. WALKER, SEARS COOK, 1805-
1853. Researches concerning
the periodical meteors of August
and November. pp.87-140. tables.
(n.s.v.8, 1843.)

1279. WILLIAMS, S[AMUEL?].
Magnetic observations, made
at the University of Cambridge
(Massachusetts) in the year 1785.
p.115. (v.3, 1793.)

(See No. 1282)

Figure 5. *Dr. Chamberlin's Electrical Bank of Health.*

MISCELLANEOUS

1280. BRUNO,
Essai metaphysique, physique
et phisiologique, relativement à
la découverte de M. Mesmer; 1786 à
St. Germain en Laye.
Bound manuscript. [20], 250p. in
French. 9-1/4 x 7".
APS

1281. CARTE MAGNÉTIQUE des deux
hémisphères.
Map. 47.3 x 92.5 cm. n.p., [ca.
1790].
APS

1282. DR. CHAMBERLIN'S electrical
bank of health...will pay
three dollars in electrical vitality...
Broadside. 3-1/4 x 7-1/2".
illus. New York, n.d.
APS

1283. CHURCHMAN, JOHN, 1753-1805.
To George Washington...this
magnetic atlas or variation chart
is humbly inscribed...
Map. 60 x 62cm. Philadelphia,
1790. Hand colored.
APS

see also: 118-20

1284. DUMAY, JULIANUS MARIA.
Redemptoris typo. Tentamina
de electricitate.
Broadside. 1p. Paris, 1779.
Dedicated to Benjamin Franklin.
APS

1285. DUNBAR, JOHN R   W   .
Letter to Mr. Robinson; Win-
chester, [Va.], June 27, 1833.
Broadside. 3p. and add. and end.
[1833].
APS

1286. ELETTROLOGIA.
Ms. notebook. [Italy, ca.
1770].
APS

1286a. EVANS, LEWIS, ca. 1700-
1756. Map of Pensilvania,
New-Jersey, New-York, and the three
Delaware counties.
Map. 19-1/4 x 25-1/4". colored.
[Philadelphia], 1749.
Contains legends about origin of
storms, lightning, electricity,
etc.
Inscription: To Dr. John Mitchel
from Mr. B. Franklin.
APS

1287. HALLEY,EDMUND, 1656-1742.
Nova & accuratissima totius
terrarum orbis tabula nautica
variationum magneticarum index
juxta observationes anno 1700 habi-
tas constructa.
Map. 47 x 113.8cm. London, 1870.
Photolithograph reproduction of
copy in Library of British Museum.
Autograph: G. B. Airy.
APS

1288. KINNERSLEY, EBENEZER, 1711-
1778. Notice is hereby given
to the curious, that at the court
house, in the council chamber, is
now to be exhibited, and continued
from day to day, for a week or two,
a course of experiments on the
newly discovered electrical fire.
Broadside, 10 x 13-1/2". Newport,
1752. Photostat.
APS

see also: 284

1289. MAGNETO ELECTRISCHE machine.
Broadside. 22 x 32cm.

[Haarlem?, 1840s].
    APS

1290. MOUNTAINE, WILLIAM, fl. 1756.
    A correct chart of the ter-
raqueous globe, on which are
described lines shewing the variation of the magnetic needle in the
most frequented seas, originally
composed in the year 1700, by...
Edmund Halley; Renewed by William
Mountaine & James Dodson F.R.S. according to observations made about
the year 1756, and now published
with all the necessary emendations.
Map. 50.8 x 120cm. London, 1794.
    APS

1291. MULHERN, EDWARD.
    Original dissertation on the
doctrine & principles of magnetism
&c. By Edward Mulhern. Philadelphia 12th Street, September 1st
1829.
Manuscript book. 41p.
    APS

1292. QUIMBY, A   M
    Lightning conductors...con-
structed & improved by A. M. Quimby
& furnished & erected under his
superintendence...
Broadside. 35.5 x 50cm. [New
York, 1841?].
    APS

1293. WAVRAN, C   L   B   .
    Essai de physique, présenté
à...Docteur Franklin...
Manuscript. 396p. 1pl. n.d.
From Benjamin Franklin's library.
    APS

1294. WOODWARD, W   H   .
    Lecture on the electro-
magnetic telegraph...March 3, 1846
...A working model will be ex-
hibited.
Broadside, 11-3/4 x 12-1/2".
West Chester, Pa., 1846.
    APS

1295. YEATES, THOMAS, 1768-1839.
    Chart of the variation of the
magnetic needle, for all the known

seas comprehended within sixty
degrees of latitude north and
south: with a new and accurate de-
lineation of the magnetic meridians,
accompanied with suitable remarks
and illustrations...Drawn & en-
graved by J. Walker.
Map. 50.5 x 120cm. [London],
1817.
    APS

1296. DEARBORN, BENJAMIN, 1754-1838.
    To the President and members
of the American Academy of Arts and
sciences. Gentlemen, As that season
of the year is approaching, in which
a general apprehension of danger from
lightning is most excited...
Broadside, 25.5 x 21.5cm. Boston,
1807.
    CP

APPENDIX: BOOKS FROM BENJAMIN FRANKLIN'S LIBRARY AND NOTES ON PROVENANCE

This checklist includes 71 volumes which were owned by Franklin, 49 on the subject of electricity or magnetism, 22 on animal magnetism. This number comprises the majority of works on these subjects which Franklin is known to have had in his collection. Sixteen additional titles or copies are held by other libraries or individuals, mostly outside the Philadelphia area.[1]

Relatively few of the other volumes in this checklist have such illustrious provenance and there are no other individuals whose libraries are so well represented. A few other names of especial interest include Joseph Priestley (5 items), James Logan (2), Benjamin Wilson (1), Benjamin Rush (1), Benjamin Smith Barton (1), and Henry Cavendish (1). These and selected other associations are noted within the entries in the main list.

Checklist numbers of volumes which Franklin owned:

> 7, 8, 11, 12, 15, 29, 31, 33, 36, 37, 48, 55, 61, 75,
> 121, 152, 153, 154, 155, 176, 191, 211, 253, 254, 272,
> 276, 301, 308, 309, 337, 350, 368, 385, 386, 436, 466,
> 515, 521, 523, 544, 545, 550, 559, 560, 561, 567, 570,
> 572, 1038, 1041, 1042, 1048, 1052, 1054, 1058, 1059,
> 1060, 1062, 1066, 1068, 1069, 1071, 1077, 1082, 1098,
> 1099, 1103, 1108, 1114, 1130, 1293.

1. This information was gleaned from a file compiled by Edwin Wolf 2nd of the Library Company of Philadelphia, of all books from Franklin's library which have been located to date.

SUBJECT AND AUTHOR INDEX

Abbot, Joel, 582
Absonus, Valentine, 1031
Adams, George, 1-6
Adanson, Michel, 7
Advertisements, 661, 757, 1282,
    1288, 1292, 1294
Aepinus, François Ulric Theodor,
    8, 507
Ahlwardt, Peter, 9
Aikin, William E.A., 584
Airy, Sir George Biddell, 660,
    749
Aldini, Giovanni, 195, 197,
    585-88
Alencé, Joachim d', 10
Alibert, Jean Louis, 589
Allen, Zachariah, 1252
Almon, William Bruce, 590
Amoretti, Carlo, 591a
Ampère, André Marie, 592-95, 760
Andry, Charles Louis François, 11
Animal magnetism, 1030-1242 and
    297, 485, 507, 512, 591a, 605,
    690, 808a, 898, 905, 916-18,
    1280
Animal magnetism - Bibliography,
    373, 1206
Antheaulme, 12, 366
Antinori, Vincenzio, 534
Arella, see Carnevale-Arella
Aschlund, Arent, 596
Ashburner, John, 917
Atlantic and Ohio Telegraph Co.,
    597
Atlantic, Lake, and Mississippi
    Telegraph Co., 598
Atmospheric electricity, 25, 29,
    33, 39 - 40, 59, 139, 194, 223,
    233, 237, 274, 290, 300, 347,
    416, 454, 466, 473, 523, 527,
    693, 798-99, 809, 825, 864.
    see also, Auroras, Comets,
    Lightning, Meteors, Thunder,
    Thunderstorms
Augustin, Friedrich Ludwig, 599-
    600
Auroras, 128-29, 337, 371, 558,
    560, 575, 601, 766, 975
Azais, Pierre Hyacinthe, 1132

Baader, Franz von, 602
Baccelli, Liberato Giovanni, 603

Bache, Alexander Dallas, 743,
    1243-46
Bachhoffner, George Henry, 604
Bachoué, see Lostalot-Bachoué
Back, Sir George, 659
Bagg, Joseph H., 605
Bailly, Jean Sylvain, 1059-61, 1067,
    1070
Bain, Alexander, 724
Bain, Sir William, 606
Balsamo, Giuseppe, 1085
Bammacarus, Nicolaus, 13
Bapst, F.G., 1132
Barbaret, Denis, 14
Barbarin, Chevalier de, 1033
Barbeguière, Jean Baptiste, 1034
Barbeu-Dubourg, 181
Barbier de Tinan, 15-16, 520
Barker, Francis, 17
Barletti, Carlo, 18-21
Barlow, James, 607
Barlow, Peter, 608-09
Barlow, William, 22, 461
Barneveld, Willem van, 23
Baronio, Giuseppe, 610
Barth, George H., 1133
Batteries, 66, 390, 534, 610,
    661, 781, 903, 986, 988, 1011,
    1022, 1257-58
Bauer, Fulgentius, 24-25
Bazin, Gilles Augustin, 26-27
Beccaria, Giovanni Battista, 20,
    28-40, 146, 510, 740
Beck, Dominikus, 41
Becket, John Brice, 42
Becquerel, Antoine César, 611-16
Becquerel, Edmond, 613
Beecher, William H., 1134, 1190
Beek, Albert van, 617
Beez, Martin, 618
Belcher, William, 43
Belgrado, Jacopo, 44
Bell, John (Reverend), 1035-36
Bell, John (1763-1820), 1037
Bell, John (1796-1872), 1135
Bella, João Antonio dalla, 45
Belli, Giuseppe, 619
Beltrami, Paolo, 620-22
Bennet, Abraham, 46
Béraud, Laurent, 47-49
Berdoe, Marmaduke, 50
Bergasse, Nicolas, 1038-39

Berna, Didier-Jules, 1136
Bertholon, Pierre, 51-57
Bertrand, Alexandre Jacques
    François, 1137
Berzelius, Jöns Jakob, friherre,
    623-28
Betti, Luigi, 58
Beudant, François Sulpice, 629
Beyer, 630
Bianchi, Tommaso, 631
Bianchini, Giovanni Fortunato, 59
Bina, Andrea, 60
Biographies, 589, 631, 887, 892,
    937
Biot, Jean Baptiste, 265, 613,
    632-35, 725
Birch, John, 4, 636
Bird, Golding, 637
Blagden, Sir Charles, 61
Blumenbach, Johann Friedrich, 62
Böckmann, Johann Lorenz, 63-64
Boetius de Boodt, Anselmus, 65
Boggia, Joseph, 388
Bohnenberger, Gottlieb Christian,
    66-69
Bombay Government Observatory, 638
Bompass, Charles Carpenter, 639
Bonci Casuccini, Alessandro, 70
Bonci Casuccini, Angelo, 70
Bond, Henry, 71
Bonnefoy, Jean Baptiste, 72,
    1040-41
Boodt, see Boetius de Boodt
Borgognini, Antonio Maria, 73
Bose, Georg Mathias, 74-77
Bostock, John, 640
Botti, Giovanni Battista, 138
Bouguer, Pierre, 78
Boulanger, Nicolas Antoine, 79
Bourzeis, Jacques Amable de, 1042
Bouvier, Marie André Joseph, 1043
Boyle, Robert, 80-84
Bragadin, Francesco Maria, 85-86
Braid, James, 1138
Brandely, A., 641
Bremner, James, 642
Bressy, Joseph, 87-88
Brewster, Sir David, 643, 923
Brewster, George, 644
Brisson, Mathurin Jacques, 89-90
British Association for the
    Advancement of Science, 645
Brook, Abraham, 91
Brossaud, Émile, 646
Broun, John Allan, 647, 858

Brown, Thomas (of Troy), 648-49
Browne, Sir Thomas, 92
Bruno, 1280
Bruno, de, 93
Brussels. Observatoire Royale,
    913
Bryant, William, 1247
Brydone, Patrick, 499
Buffon, George Louis Leclerc,
    comte de, 650
Burdin, Charles, 1139
Burnett, Charles Mountford, 1140
Burton, Frances Barbara, 1141
Bush, George, 1142
Bywater, John, 651-52

Cabanis, Pierre Jean Georges,
    1044
Cabeo, Niccolo, 94
Cagliostro, Alexandro, conte, 1085
Calandrelli, Giuseppe, 95
Caldwell, Charles, 62, 1143
Cambry, Jacques, 1045
Canton, John, 96, 178, 366
Carnevale-Arella, Antonio, 653
Carpue, Joseph Constantine, 654
Carradori, Gioachino, 655
Caullet de Veaumorel, 383, 1086
Caustic, Christopher, pseud.,
    see Fessenden
Cavallo, Tiberius, 97-112
Cazeles, see Masars de Cazeles
Ceppi, Luigi Antonio, 113
Chamberlain, Dr., 1282
Channing, William Francis, 656
Chappe, Ignace Urbain Jean, 657
Chardel, Casimir Marie Marcellin
    Pierre Célestin, 1144
Charpignon, Jules, 1145
Charpignon, Louis Joseph Jules,
    see Charpignon, Jules
Chastellux, François Jean, marquis
    de, 1038
Chastenet, see Puységur
Chernak, Ladislaus, 114
Chevalier, Charles Louis, 658
Chigi, Alessandro, 116-17
Christie, Samuel Hunter, 659-60
Christmas, Henry, 1146
Churchman, John, 118-20, 1283
Cigna, Giovanni Francesco, 121
Coad, Patrick, 661
Colden, David, 178
Collyer, Robert H., 1147
Colquhoun, John Campbell, 1148-50,
    1188

Comets, 418

Commun, Joseph du, 1151

Compass, 22, 78, 120, 189, 231,
    261, 275, 317, 333, 366-68,
    471, 540-43, 575, 596, 606,
    617, 678, 690, 758, 813, 856,
    914, 923, 950, 981, 1256, 1273

Constantini, Giuseppe, 122

Cooper, C. Campbell, 662-63

Cooper, M., 123

Cornacchini, Pietro, 124

Cosnier, see Ledru

Costantini, see Constantini

Cotton, Charles, 1152

Coudret, J.F., 664

Courtenay, Edward Henry, 1244

Creve, Johann Caspar Ignaz Anton,
    125

Croker, Temple Henry, 126

Crumpe, (Miss) G.S., 1153

Crusell, Gustav Samuel, 665

Cumming, James, 683

Cunningham, Peter Miller, 666

Cuthbertson, John, 667-68

Cuypers, C., 127

Dalancé, see Alencé

Dalibard, Thomas François, 171-72

Dalton, John, 128-29

Dampierre, Antoine Esmonin,
    marquis de, 1032, 1047-48

Darwin, Erasmus, 130-32

Davenport, Thomas, 669, 750, 959

Davies, John, 670

Davis, Daniel, jr., 671-74

Davy, Sir Humphry, 393, 675-79, 887

Dearborn, Benjamin, 1296

De Fremery, Nicolaus Cornelius, 133

De Kramer, Antonio, 680

De La Rive, see La Rive

Delaunay, Claude Veau, see Veau
    De Launay

Delesse, Achille Ernest Oscar
    Joseph, 681-82

Deleuze, Joseph Phillippe
    François, 1154-59

Demonferrand, Jean Baptiste
    Firmin, 683-84

Dermogine, Epugispe, 685

Desaguliers, Jean Théophile, 134-35

Deslon, see Eslon

Desmarest, Nicolas, 244

Despretz, César Mansuète, 686

Devillers, Charles, 1049

Díaz de Gamarra y Dávolos, Juan
    Benito, 136

Digby, Sir Kenelm, 137

Divining rod, 591a

Divisch, Procopius, 139

Dods, John Bovee, 1160-61

Dollond, Peter, 687

Domin, Joseph Francis, 140

Donndorff, Johann August, 141-42

Donovan, Michael, 688-89

Doyle, George S., 690

Draper, John William, 690a

Dubois, Fred., 1139

Dubois-Reymond, Emil Heinrich,
    691-92

Dufay, Charles François de
    Cisternay, 143

Duhamel, Jean Baptiste, 144, 366

Dumay, Julianus Maria, 1284

Dunbar, John R.W., 1285

Duncan, George W., 1162

Dupau, Jean Amédée, 1163-64

Du Potet de Sennevoy, Jules,
    baron, 1165-67

Duprez, François Joseph Ferdinand,
    693

Dutour, Étienne François, 145

Eandi, Giuseppe Antonio Francesco
    Girolamo, 146

Eberhard, Christoph, 147

Edwards, William Frédéric, 694-96

Ehrmann, Friedrich Ludwig, 148

Electrical apparatus, 4, 46, 63,
    66-69, 89, 91, 127, 152-54,
    159, 162-63, 211, 223-24, 226-
    27, 235, 243, 245-46, 272, 302,
    308, 336, 354-56, 371, 382-84,
    387, 391-92, 415, 436, 460,
    546, 552, 554, 565, 576-78,
    610, 641, 654, 669, 674, 680,
    687, 698, 717, 727, 750, 766,
    769, 771, 801, 844-45, 849,
    853, 874, 959, 983, 1002, 1009-
    10, 1015, 1019, 1253-54, 1289

Electrical clocks, 724, 909, 948

Electrical currents, 594, 604,
    787, 873, 879, 903, 1013

Electrical lamps, 140, 148

Electrical musical instruments, 298

Electrical organs in fishes, 691,
    739, 1248. see also, Torpedo

Electricity - Bibliography, 1002

Electricity
    Effect on plants, 52, 57, 274

Electricity
History, 171, 340, 365, 395
440, 442-47, 482-83, 491-92,
584, 599, 640, 652-53, 1010
Electrochemistry, 139, 393, 611-
13, 624-28, 641, 726, 753,
878, 920, 953, 963-64, 978,
989
Electrodynamics, 593-95, 683-84,
715-16, 737, 795, 894, 979,
1007, 1259-60
Electromagnetic apparatus, 732
Electromagnetism, 592, 594, 603-
04, 608-09, 666, 669, 671-73,
680, 698, 702-03, 706-08, 710,
714, 717, 733, 747, 750, 760,
767, 779, 853-54, 870-71, 874,
876-77, 899, 902, 926, 959, 974,
983, 989, 1007, 1019, 1021, 1289
Electrometallurgy, 658, 713, 802-
03, 830, 949, 955-56, 966-67,
969-70, 977-78, 1005
Electroplating, 953, 986
Electrostatics, 746
Electrotherapeutics (General), 2,
4, 11, 18, 23, 42, 53-54, 63,
72, 98, 100, 104-05, 124, 161-63,
200, 211-15, 230, 234, 306, 319-
20, 344, 360-63, 382-83, 424,
428, 459, 469-70, 479, 486, 500,
508, 513, 526, 530, 564, 581,
590, 599-600, 636-37, 646, 648-
49, 654, 656, 661, 664-65, 668,
674, 689, 727, 732-34, 756-57,
789, 801, 818-19, 833, 846, 863,
885, 895-97, 905, 910, 919, 946,
957-58, 965, 974, 1002, 1008-09,
1282
Electrotherapeutics (Specific
ailments)
Epilepsy, 306
Gout, 946, 1009
Insanity, 124
Nervous diseases, 306, 646, 897,
905, 919, 1009
Paralysis, 315, 469-70, 564,
733, 833, 946
Parturition, 961
Rheumatism, 564, 646, 732-33,
946, 1009
Yellow fever, 957
Electrotyping, 671, 886, 1004,
1014
Elias, Pieter, 698, 1289
Elice, Ferdinando, 699-701

Elliotson, John, 1168
Epp, Franz Xavier, 149
Erxleben, Johann Christian
Polykarp, 150
Eschenmayer, Carl Adolph von, 1239
Esdaile, James, 1169-70
Eslon, Charles d', 1051-54
Esprémesnil, Jean Jacques Duval d',
1055-56, 1114-15
Esser, Ferdinand, 151
Étienne, d', 152-54
Euler, Johann Albrecht, 155-56
Euler, Leonhard, 157
Evans, Lewis, 1286a
Everett, Jesse, 648-49
Exley, Thomas, 702
Eydam, Immanuel, 703

Fabré-Palaprat, Bernard-Raymond, 818
Fansher, Sylvanus, 704
Faraday, Michael, 705-12, 915
Fardely, William, 713
Farrar, John, 714-16
Faulwetter, Carl Alexander, 158
Faure, Giovanni Battista, 159
Fechner, Gustav Theodor, 717-18
Felbiger, Johann Ignatz von, 160
Feller, Christian Gotthold, 161
Ferguson, James, 162-63
Fessenden, Thomas Green, 719-23
Fillassier, Alfred, 1172
Finlaison, John, 724
Fischer, Ernst Gottfried, 725
Fischer, Nicolaus Wolfgang, 726
Fisher, George Thomas, 727
Flagg, Henry Collins, 1248
Florence. Accademia del Cimento,
164
Fludd, Robert, 165
Foissac, Dr., 1159
Foissac, Pierre, 1173
Follini, Giorgio, 167
Fonda, Girolamo Maria, 168
Fontette-Sommery, comte de, 1058
Forbes, James David, 728-29
Forbes, Sir John, 1174-75
Forbes, Robert Benjamin, 784
Forman, J.G., 1241
Fournel, Jean François, 1176
Fowler, Richard, 169
Fox, Robert Were, 730
France. Commission Chargée de
l'Examen du Magnétisme Animal,
1059-72
Franklin, Benjamin, 170-84, 519,

1059-61, 1064-67, 1070-72, 1249-50, 1286

Freke, John, 185-89

French, Benjamin Franklin, 1211

Friedlander, Michael, 1177

Friese, Robert, 731

Frisi, Paolo, 190-92

Froriep, Robert, 732-33

Galart de Montjoie, see Montjoie

Gale, Thaddeus, 734

Galisteo, Juan, 193

Galitzin, Dmitri Alexievitch, prince, 194

Galvani, Luigi, 195-99, 589, 740

Galvanism - History, 639, 655, 688, 863, 954, 990

Gardini, Giuseppe Francesco, 200-01

Gauss, Karl Friedrich, 735-36

Gavarret, Louis Dominique Jules, 737

Gay-Lussac, Joseph Louis, 738

Gemminger, Max, 739

Gherardi, Silvestro, 740

Gianni, Francesco, 741

Gilbert, William, 202-04

Gilliss, James Melvin, 742

Girard College, Philadelphia, 743

Girardin, Dr., 1073-74

Glanvill, Joseph, 205-06

Gnillius, Johannis Henricus, 207

Goclenius, Rudolph, 208

Gonfigliacchi, Pietro, 537

Gordon, Andrew, 209-10

Gorgeret, Philibert Marie Euzèbe, 1178

Gorham, John, 744

Graham, James, 211-15

Graham, James Duncan, 1251

Grandami, Jacques, 216

Grandvoinet, J.A., 1179

Great Britain. Ordnance & Admiralty, Dept. of, 745

Green, George, 746

Green, Jacob, 747-48

Greenwich. Royal Observatory, 749

Gregory, George, 217

Gregory, William, 916, 918

Griglietta, C., 750

Grimelli, Geminiano, 740, 751

Grimes, James Stanley, 1179a

Grimston, Henry, 752

Griselini, Francesco, 218

Gross, Johann Friedrich, 219-20

Grothuss, Theodor, freiherr von, 753

Grund, Francis Joseph, 754

Guden, Philipp Peter, 221

Guericke, Otto von, 222

Gütle, Johann Conrad, 223-27

Gusserow, Carl August, 755

Guyot, Edmée-Gilles, 228

Haddock, Joseph W., 1180

Hale, Sir Matthew, 229

Haliday, William, 230

Hall, Richard Willmott, 756

Halley, Edmund, 1287

Halse, William Hooper, 757

Hamard, Charles Pierre Guillaume, 1181

Hamberger, Georg Erhard, 231

Hansteen, Christophe, 758, 942

Hare, Robert, 759-82, 1252-55

Harrington, Robert, 232

Harris, William Snow, 783-86

Harsu, Jacques de, 1075

Hartmann, Johann Friedrich, 233-37

Hartshorn, Thomas C., 1156-59

Hassler, Ferdinand Rudolph, 1256

Haüy, René Just, 238-42

Hauksbee, Francis, 243-46

Hausen, Christian August, 247-48

Hays, Isaac, 1182

Hazard, Ebenezer, 748

Hell, Maximilian, 249

Helmont, Johann Baptist van, 250

Hemmer, Johann Jakob, 251-52

Hénin de Cuvillers, Étienne Felix, baron d', 1163, 1183-84, 1239

Henley, William, 253-54, 436, 466

Henrici, Friedrich Christoph, 787

Henry, Joseph, 988, 1257-60

Herbert, Joseph, Edler von, 255-56

Héricart de Thury, Louis Étienne François, vicomte, 788

Hervier, Charles, le père, 1077-77a

Hickmann, Johann N., 789

Higgins, William, 790

Higgins, William Mullinger, 791-92

Highton, Edward, 793

Hirschel, Anastasius, 794

Hitt, J.D., 795

Hoadly, Benjamin 257-58

Hobart. Magnetical and meteorological observatory, 796

Hodgkin, Thomas, 695-96

Höll, see Hell

Hoffmann, Carl Ferdinand, 259

Hooper, William, 260
Hopkins, Evan, 797
Hopkins, George F., 798
Hopkinson, Francis, 1275
Howard, Edward, of Berks., 261
Howard, Luke, 696, 799
Hübner, Lorenz, 507
Humboldt, Alexander, freiherr von, 262-67, 660
Hunter, John, 268
Husson, 1137, 1214

Imhof, Maximus, 269-70
Inchbald, Mrs. Elizabeth (Simpson), 1186-87
Ingenhousz, Johannes, 271-78
Institut de France. Academie des sciences, 1188
Irvine, Christopher, 279
Izarn, Joseph, 800-01

Jackson, Charles T., 811
Jacobi, Moritz Hermann, 802-03
Jacquet de Malzet, Louis Sébastien, 280
Jadelot, Jean François Nicolas, 262-63
Jallabert, Jean, 281-82
Johnson, Charles P., 1189
Johnson, Edward John, 804
Johnson, Moses, 805
Johnson, Walter Rogers, 806
Jones, Henry, 1190
Jones, John, 1276
Jones, William, Rev., 283
Joyce, Jeremiah, 807-08
Józwik, Albert, 1191
Judel, René François, 808a
Jussieu, Antoine Laurent, 1062-63

Kaemtz, Ludwig Friedrich, 809
Karsten, Karl Johann Bernhard, 810
Kendall, Amos, 811-12
Kinnersley, Ebenezer, 284, 1288
Kircher, Athanasius, 285-88
Kirchhof, Nicolaus Anton Johann, 289
Kirchvogel, Andreas Bernhard, 290
Kites, 416
Klaproth, Heinrich Julius von, 813
Klinger, J.A., 1193
Klügel, Georg Simon, 291
Kluge, Carl Alexander Ferdinand, 1194-95
Knight, Gowin, 292-94

Knobloch, M., 814
Koelle, August, 815
Koestlin, Karl Heinrich, 295
Kratzenstein, Christianus Gott-lieb, 296
Krünitz, Johann George, 440
Kümpel, Johannes Andreas Fridericus, 297
Kupffer, Adolphe Theodor, 816, 936

La Beaume, Michael, 818-19
La Borde, Jean Baptiste de, 298
Lacépède, Bernard Germain Étienne de La Ville sur Illon, comte de, 299
Lafonde, see Sigaud-Lafond
Lafontaine, Ch., 1196
L'Aming, Richard, 820
Lamont, Johann von, 821
Lampadius, Wilhelm August, 300
Landriani, Marsiglio, 301
Lang, William, 1197
Langenbucher, Jakob, 302-03
Languth, Carl Julius, 822
La Perrière de Roiffé, Jacques Charles François de, 304
Lapostolle, Alexandre, 823
Lardner, Dionysius, 824-25
La Rive, Arthur Auguste de, 1026
Laugier, Esprit-Michel, 1077
Lavoisier, Antoine Laurent, 305
Le Bouvier-Desmortiers, Urban René Thomas, 826
Ledru, Nicolas Philippe, 306
Le Duc, W.G., 1179a
Lee, Edwin, 1198-99
Léger, Theodoro, 1200
Leibl, 827
Leitch, John M., 828
Leithead, William, 829
Le Monnier, Pierre Charles, 307
Lerebours, Noël Marie Paymal, 830
Le Roy, Jean-Baptiste, 308-09
Le Yerrier, M., 832
Lichtenberg, Georg Christoph, 150, 310
Lichtenberg, Ludwig Christian, 311
Liebig, Justus, 867
Lightning, 9, 14, 18, 28, 46, 55, 58, 61, 116-17, 122, 138, 149, 193, 233, 252-53, 259, 284, 291, 303, 308, 311, 321, 330-31, 334, 338-39, 431, 433, 454, 457, 459, 531-32, 539, 574, 602, 670, 784, 788, 1270, 1275-76, 1286a, 1296

Lightning conductors, 15-16, 45, 55, 56, 64, 85-86, 95, 115, 133, 151, 160, 162-63, 168, 181, 194, 220-21, 223, 235, 251, 254, 269, 289, 301, 303, 309, 311, 330, 338-39, 341, 369, 386, 451, 457-58, 465, 472, 497-98, 516-21, 523, 531, 566, 568, 570, 591, 620-22, 630, 699, 701, 704, 738, 783, 823, 857, 883, 1274, 1292
Linnaeus, Carolus, 581
Lisicki, 833
Litta Biumi Resta, Carlo Matteo, 1079
Lloyd, Humphrey, 834-43, 941
Lobe, Wilhelmus, 312
Locke, John, 844-45, 1261-63
London Electrical Society, 1028-29
Longinus, see Goclenius
Loomis, Elias, 1264-67
Lorimer, John, 313
Lostalot-Bachoué, Jean Pierre, 846
Loubert, Jean Baptiste, abbé, 1201
Loughborough, John, 847
Louis, Antoine, 314-15
Lovering, Joseph, 716, 848
Lovett, Richard, 316-18
Lowndes, Francis, 319-20
Lozeran du Fesc, Louis Antoine de, 321
Luc, Jean André de, 849
Lüderus, Gerhardus, 324
Ludwig, Christian, 322-23
Lugt, Hendrik, 325-26
Lull, Ramón, 327
Lullin, Amadeus, 328
Lutzebourg, comte de, 1080-81
Luyts, Jan, 329
Luz, Johann Friedrich, 330
Lyon, John, 331-32
Lyons, L'École Vétérinaire, 1082

Mackay, Andrew, 333
Macrery, Joseph, 850
Madison, James, Bp., 1268
Maffei, Francesco Scipione, marquese de, 122, 334
Magalotti, Lorenzo, 164, 335
Magarotto, Antonio, 851
Maggiotto, Francesco, 336
Magnetic declination, 71, 78, 313, 324, 414, 506a, 736, 838, 842, 1251
Magnetic instruments, 730, 835-37, 996

Magnetic observatories, 821, 836, 931, 934-35
Magnetic Telegraph Co., 852
Magnetism - Bibliography, 373
Magnetism - Terrestrial, 78, 118-20, 265, 307, 542, 557, 607-09, 614, 645, 660, 680, 728, 735-36, 797, 834-35, 837, 839-40, 848, 864, 906, 911, 927-28, 930, 932-33, 938-44, 951, 996, 1018, 1022, 1245-46, 1281, 1283, 1287, 1290, 1295
Magnetism - Terrestrial - Observations, 638, 647, 659, 729, 742-43, 745, 749, 796, 816, 842, 858, 908, 913, 936, 945, 992, 994, 1243-44, 1261-67, 1269, 1279
Magnetism - Therapeutic use, 11, 208, 250, 279, 353, 425-26, 503, 511, 524, 719-23, 827
Magnetism of ships, 606, 609, 617
Magnets, 10, 12, 26-27, 126, 137, 229, 249, 261, 352, 366-68, 543, 822, 841, 843
Magrini, Luigi, 853-55
Mahon, see Stanhope
Mairan, Jean Jacques Dortous de, 337
Maissait, Michel, 856
Maitland, Samuel Roffey, 1202
Majocchi, Giovanni Alessandro, 622, 857
Makerstoun Observatory, 647, 858
Mako, Paul, 338-39
Mangin, abbé, 340
Marat, Jean Paul, 342-46
Marcard, Dr., 1226
Marherr, Philipp Ambrosius, 347
Mariotti, Prospero, 348
Markus, M., 665
Martin, Benjamin, 349-52
Martin, John, 1084
Martineau, Harriet, 1203-04
Martius, Johann Nicolaus, 353
Marum, Martin van, 354-59
Masars de Cazèles, 360
Matteucci, Carlo, 859-62
Mauduyt de la Varenne, Pierre Jean Étienne, 361-63
Mayer, Federico, 1017
Mazzacane, Carlo, 364
Mazzolari, Giuseppe Maria, 365
Meade, William, 863
Mesmer, Friedrich Anton, 507,

1033, 1068, 1086-94, 1216
Metcalfe, Samuel Lytler, 864
Meteors, 51, 1273, 1278
Mialle, Simon, 1206
Michell, John, 366-68
Mills, John, 369
Milner, Thomas, 370
Mitchell, John, 509
Moigno, François Napoléon Marie, 865
Montanari, Giuseppe Ignazio, 537
Montegre, Antoine François Jean de, 1207
Montjoie, Christoph Felix Louis Ventre de la Touloubre, 1095
Morgan, George Cardogan, 371
Morin, Jean, 372
Morin, Pyrame Louis, 866
Morse, Samuel Finley Breese, 831, 882
Mouillesaux, 1096
Mountaine, William, 1290
Mueller, C.H., 963
Mulhern, Edward, 1291
Müller, Johann Heinrich Jakob, 867-68
Murhard, Friedrich Wilhelm August, 373
Muscio, Gian Gaetano de, 374
Musschenbroek, Petrus van, 164, 375-81

Nairne, Edward, 61, 63, 253, 382-86
Neale, John, 387
Neckermann, Michael, 388
Needham, John Turberville, 389, 497
Negro, Salvatore dal, 390-92, 854
Newnham, William, 1208-09
Nicholson, William, 393-94
Nicollet, Joseph Nicolas, 1269
Noad, Henry Minchin, 870-72
Nobili, Leopoldo, 873-75, 892
Nollet, Jean Antoine, 314, 395-413
Norman, Robert, 414
Noya Carafa, duc de, pseud., see Adanson

Oersted, Hans Christian, 876-78
O'Gallagher, Felix, 417
Ohm, Georg Simon, 879
Oliver, Andrew, 418, 1270-71
Olmsted, Denison, 880-81
O'Rielly, Henry, 882
Orioli, Francesco, 883
Osann, Gottfried Wilhelm, 884

Ozanam, Jacques, 419

Page, John, 1272
Pallas, Emmanuel, 885
Palmer, W., 886
Palmieri, Luigi, 1022
Paris, John Ayrton, 887
Paris. Faculté de médecine, 1098-99
Park, Andrew H., 888
Park, Roswell, 889
Parthenius, Joseph Marianus, see Mazzolari
Partington, Charles Frederick, 890
Pasley, T.H., 1210
Patterson, Robert, 1273-74
de Paula Candido, Francisco, 891
Paulet, Jean Jacques, 1100-01
Paulian, Aimé Henri, 420
Pearson, John, 1102
Peart, Edward, 421-22
Pelli-Fabbroni, Giuseppe, 892
Penrose, Francis, 423-24
Pereira, Jonathan, 860
Perkins, Benjamin Douglas, 425-26
Person, Charles Cléophas, 893
Peschel, Karl Friedrich, 894
Peter, Robert, 895
Petetin, Jacques Henri Désiré, 896-97
Petiau, abbé, 1103
Peytavin, Jean-Baptiste, 898
Pfaff, Christian Heinrich, 899
Pickel, Johann Georg, 427
Piette, Jacques Edme, 900
Pigeaire, J., 1212
Pinnock, William, 901, 921
Pivati, Giovanni Francesco, 428
Plaz, Antonius Guilielmus, 429
Poe, Edgar Allan, 1211, 1212a
Poetry, 58, 73, 130, 132, 365, 467, 502, 719-23, 741, 1044
Pohl, Georg Friedrich, 902-03
Pohl, Joseph, 430
Poisson, Siméon Denis, 738
Poissonnier-Desperrieres, Antoine, 1068-69
Poli, Giuseppe Saverio, 431
Polinière, Pierre, 432
Pollock, Thomas, 904
Pomme, Pierre, 905
Poncelet, Polycarpe, 433
Pope, William, 906
Portwine, Edward, 907
Pouillet, Claude Servais Matthias, 696

Power, Henry, 434
Poyen Saint Sauveur, Charles, 1213-14
Prague. Kaiserlich Königliche Sternwarte, 908
Pressavin, Jean Baptiste, 1105
Prévost, Pierre, 435
Priestley, Joseph, 436-47, 937
Pringle, Sir John, 448
Progress, Peter, pseud., 909
Puget, Louis de, 449
de Puisaye, Charles, 910
Puységur, Armand Marie Jacques de Chastenet, marquis de, 1034, 1106-12
Puységur, Jacques Maxine Paul de Chastenet, comte de, 1108, 1113-14

Quetelet, Lambert Adolphe Jacques, 911-14
Quimby, A.M., 1292
Quintine, l'abbé de la, 450

Rabiqueau, Charles A., 451
Rackstrow, B., 452
Rancy, de, 453
Read, John, 454
Reade, Joseph, 915
Reamur, René Antoine Ferchault de, 143
Reese, David Meredith, 1215
Reichenbach, Karl Ludwig Friedrich, freiherr von, 916-18
Reimarus, Johann Albrecht Heinrich, 457-58
Retz, Noël, 459, 1119
Retz de Rochefort, Noël, 459, 1119
Retzer, Joseph Friedrich, freiherr von, 338-39
Rhubarbini de Purgandis, pseud., see Servan
Riadore, John Evans, 919
Ribright, George, 460
Ricard, J.J.A., 1216-17
Ridley, Mark, 461
Rittenhouse, David, 1272, 1275-77
Ritter, Johann Wilhelm, 462, 920
Rivoire, Antoine, 366
Roberts, George, 901, 921
Roberts, Martyn J., 922
Robison, John, 923
Rönnberg, Bernhard Heinrich, 463
Roessinger, Fréderic Louis, 924
Rogers, Henry J., 925

Roget, Peter Mark, 926
Rohault, Jacques, 464-65
Ronayne, Thomas, 466
Ross, James Clark, 927-28
Roth, Johann Joseph, 929
Royal Society of London. Committee of Physics and Meteorology, 930-35
Russia. Observatoire Physique Central Nicolas, 936
Rutt, John Towill, 937

Sabine, Sir Edward, 745, 796, 938-45, 992
St. Helena. Magnetical and Meteorological Observatory, 945
Salaville, Jean Baptiste, 1120
Sambuceti, Luigi Maria, 467
Sandby, George, 1218
Sanden, Heinrich von, 468
Sans, abbé, 469-70
Sarlandière, Jean Baptiste, 946
Sarrabat, Nicolas, 471
Saussure, Horace Bénédict de, 472-73, 527
Savi, Paolo, 862
Scarella, Giovanni Battista, 474
Scarso, Joseph, 947
Schäffer, Jakob Christian, 475-78
Schaeffer, Johann Gottlieb, 479
Schellen, Thomas Joseph Heinrich, 948
Schmidt, Christian Heinrich, 949
Scoresby, William, 950-52
Secondat, Jean Baptiste, baron de, 480-84
Segnitz, Friedrich Ludwig, 485
Selmi, Francesco, 953
Servan, Antoine Joseph Michel, 1121-23
Seyffer, Otto Ernst Julius, 954
Sguario, Eusebio, 486
Shaw, George, 955-56
Shecut, John Linnaeus Edward Whitridge, 957
Sherwood, Henry Hall, 958, 996, 1219
Sigaud-Lafond, Joseph Aignan, 487-93
Silliman, Benjamin (1779-1864), 750, 959
Silliman, Benjamin (1816-1885), 960
Simon, Claude Gabriel, 1220
Simpson, Sir John Young, 961

124

Singer, George John, 962-64
Smee, Alfred, 965-72
Smith, Francis Ormond Jonathan, 973
Smith, Gibson, 1221
Smith, Horatio, 1222
Smith, Samuel B., 974
Snow, Robert, 975
Société de Guienne, 1118
Société de l'Harmonie d'Ostende, 1033, 1094
Socin, Abel, 494
Somerville, Mrs. Mary (Fairfax), 959, 976
Sousselier de la Tour, 1124
South, Thomas, 1223
Spallanzani, Lazzaro, 495-96, 589
Sparks, 271, 276, 1253
Spencer, Thomas, 977-78
Squario, see Sguario
Stanhope, Charles, 3rd earl, 497-99
Stearns, Samuel, 1125
Steavenson, Robert, 500
Steiglehner, Coelestino, 501, 507
Steinberg, Karl, 979
Steinheil, Karl August, 980
Stevenson, William Ford, 981
Stone, William Leete, 1072, 1224-25
Strada, Famianus, 502
Strombeck, Frederic Charles, baron de, 1226
Stubbe, see Stubbs
Stubbs, Henry, 503
Sturgeon, William, 982-89, 1025
Sturm, Johann Christoph, 504
Sue, Pierre, 990
Sulzer, Johann Georg, 505
Susquehanna River and North and West Branch Telegraph Co., 583
Sutherland, La Roy, see 1240
Swammerdam, Jan, 506
Swedenborg, Emanuel, 506a, 1142
Swinden, Jan Hendrik van, 507
Symes, Richard, 508
Symmer, Robert, 509
Sympathy, 137, 205-06, 279, 502, 562

Tana, Agostino, 510
Tardy de Montravel, A.A., 1126-27
Telegraph, 724, 793, 805, 811-12, 831, 855, 865, 882, 888, 907, 909, 925, 948, 973, 980, 998-1000, 1294
Telegraph - France, 832
Telegraph - History, 657

Telegraph - U.S., 583, 597-98, 828, 847, 852, 995, 997-1000, 1006
Teste, Alphonse, 1227-31
Thermoelectricity, 1020
Thillaye, 962
Thomson, Thomas, 991
Thouret, Michel Augustin, 11, 1043, 1072, 1115, 1128-29
Thouvenel, Pierre, 512-14
Thunder, 431, 433
Thunderstorms, 14, 96, 235, 251-52, 305, 338-39, 499, 783, 798, 859, 1270
Tissart de Rouvres, Jacques Louis Noël, marquis de, 1130
Titius, Johann Daniel, 515
Toaldo, Giuseppe, 85, 516-20, 527
Toderini, Giambattista, 521
Tonna, Mrs. Charlotte Elizabeth (Browne), 1232
Tornadoes, 766, 1252, 1255, 1271
Toronto. Magnetical and Meteorological Observatory, 992
Torpedo, 268, 448, 495, 544-45, 862, 1247. see also, Electric organs in fishes
Tourmaline, 7, 569, 580
Townshend, Chauncey Hare, 1233-36
Tractors, metallic, 425-26, 564, 719-23, 752
Tressan, Louis Elisabeth de la Vergne, comte de, 522
Turini, Pietro, 523

U.S. - Naval Observatory, 994
U.S. - Treasury Department, 995
U.S. 25th Congress. 2nd Session. Senate. Committee on Naval Affairs, 996
U.S. 28th Congress. 2nd Session. House, 997
Unzer, Johann Christoph, 524

Vail, Alfred, 998-1000
Vallemont, Pierre Le Lorrain de, 525
Valli, Eusebio, 526
Vanuxem, Lardner, 1001
Vasquez y Morales, Joseph, 395
Vassalli-Eandi, Antonio Maria, 527-28
Vaughan, John, 529
Veau de Launay, Claude Jean, 1002
Vène, A., 1003
Ventre de la Touloubre, see Montjoie

Veratti, Giuseppe, 530
Viacinna, Carlo, 531
Villeneuve, Olivier de, 532
Vollmar, Johannes Guillielmus, 533
Volta, Alessandro Giuseppe Antonio
    Anastasio, conte, 534-38, 631

Wagner, Godofredus, 539
Wakeley, Andrew, 540
Walckiers, 89
Walker, Charles Vincent, 809, 824,
    1004, 1026
Walker, Ralph, 541-43
Walker, Sears Cook, 1278
Wall, Arthur, 1005
Walsh, John, 544-45
Washington and New Orleans Tele-
    graph Co., 1006
Watkins, Francis, 546
Watkins, Francis, 1007
Watson, Sir William, 413, 547-51
Wavran, C.L.B., 1293
Weber, Georges P.F., 1008
Weber, Joseph, 552-56, 1027
Weber, Wilhelm Edward, 736
Weigel, Christian Ehrenfried, 345
Wesermann, H.M., 1237
Wetzler, Johann Evangelist, 1009
Whiston, William, 557
Whitney, S. Albert, 690
Widenmann, Johann Friedrich Wil-
    helm, 220
Wiedeburg, Johann Ernst Basilius,
    558
Wienholt, Arnold, 1150
Wightman, Joseph Milner, 1010
Wilcke, Johann Carl, 176, 559-61
Wilkins, John, bp., 562
Wilkinson, Charles Henry, 563-65
Williams, S., 1279
Wilson, Benjamin, 257-58, 566-72
Windler a Stortewagen, Peter
    Johann, 573
Winkler, Johann Heinrich, 574-79
Woodward, W.H., 1294

Yeates, Thomas, 1295

Zallinger zum Thurn, Franz Sera-
    phim, 580
Zamboni, Giuseppe, 1011
Zantedeschi, Francesco, 1012-23
Zetzell, Per, 581